THE SYNTONIC PRINCIPLE

Its Relation to Health and Ocular Problems

THE SYNTONIC PRINCIPLE

Its Relation to Health and Ocular Problems

By

HARRY RILEY SPITLER, D.O.S., M.D., M.S., Ph.D.

Formerly Clinician Macfadden Sanatorium, Battle Creek, Mich.; Physician-in-charge Crab Orchard Sanatorium, Crab Orchard, Ky.; Past President State Board of Optometry, Ohio; Past First Vice-President American Optometric Association; Accredited teacher of Mechanotherapy and physical therapies since 1925 by Ohio State Medical Board; Past Dean Department of Mechanotherapy Metropolitan College; Dean Central States College of Physiatrics; Fellow American Academy of Optometry; Fellow College of Syntonic Optometry.

RESOURCE *Publications* · Eugene, Oregon

Resource Publications
A division of Wipf and Stock Publishers
199 W 8th Ave, Suite 3
Eugene, OR 97401

The Syntonic Principle
Its Relation to Health and Ocular Problems
By Spitler, Harry Riley, DOS, MD, MS, PhD
ISBN 13: 978-1-6097-746-3
Publication date 9/1/2011
Previously published by College of Syntonics, 1941

Dedicated to

FREDRIC A. WOLL, Ph.D.,

whose example, kindly and gentle counsel, fatherly interest and scientific inquisitiveness have proved a constant inspiration to the author.

PREFACE

In the preparation of the following presentation the author will endeavor to present certain experimentally determined facts and some clinical evidence in support of his thesis that radiant energy in the photic range of the electromagnetic spectrum plays an important part in the development of, and the health of human beings.

Further, effort will be made to make it clear that the frequency of light in the photic range, and not its intensity, when incident into the eyes of man, and most animals, has a determining effect upon his growth, development, and functions, physiologic, ocular, and psychic. No effort will be made to give light frequency prescriptions for altering responses because such attempts must be predicated upon a full understanding of each case.

The author freely admits that in the selection of material for the many citations in the text he has been primarily motivated to select material which would complete the "mosaic" of the syntonic concept, although many citations which appear contradictory upon first thought were also included because they illustrated a point to be discussed at the time, or later in the manuscript.

The author is fully aware that there will be criticism of his material, perhaps his conclusions, and he will welcome constructive criticism by informed and qualified critics, yet in extenuation of his conclusions he wishes to remind readers that the facts speak for themselves, regardless of his attempts at their explanation.

In an addendum will be found a small series of clinical case reports illustrative of the application of the principles involved. It has well been said that clinical results prove nothing. That is true. Yet in all of the biological professions, the clinical test is the final test of effectivity. For that reason a clinical effectivity tabulation of over three thousand cases has been included in the appendix.

Grateful acknowledgment is made by the author for the stimulus to continued effort in the preparation of this work to the gentle and sometimes not so gentle prodding of those having his interest at heart, and for the able assistance in preparing the manuscript given by his secretary, Edith Donohoe; also for the drawings and work on the index by Evelyn Kuns.

Confession is further made that the author undoubtedly would be able to find passages which he might well rewrite to make them meet possible criticism by others as well as his own. It is in the spirit of John Henry Newman, expressed in the following that motivates release of the material at this time just as it is:

"Nothing would be done at all, if a man waited 'til he could do it so well that no one could find fault with it."

H. R. S.

Eaton, Ohio
April 25, 1939

Publisher's Statement

This book is in part an abridgement and in part an amplification of a thesis in part fulfillment of academic requirements for the Doctor of Philosophy degree.

TABLE OF CONTENTS

PART I

THE PROBLEM

CHAPTER I

It would seem that the logical order of approach to a discussion of the subject matter of this research dictates that the basic terms be defined as the first step.

Webster's International Dictionary defines *syntonic* and *syntony* in terms of physics. The origin of the word is stated to have been from the Greek *syn* meaning "like or the same"; and *tonos* variously translated as "tone, pitch, tension, and tonicity." *Webster* then defines the word *syntony* as "the state of being adjusted to a certain wave length; agreement or tuning between the time period of an apparatus; agreement."

Syntonic is defined as "the state of syntony between transmitting and receiving apparatus." Syntonizer is defined as "one that syntonizes or produces syntony; one who tunes; a device for tuning such as is used in wireless telegraphy."

The foregoing definitions obviously do not apply to physiologic processes, but are limited to mechanisms which may be caused to have equal tensions and masses in order that they may be tuned together, i.e., to the same frequency or pitch. In physiology no such fixed frequency exists unless perhaps the Berger rhythm might be so considered. This rhythm will be discussed at length later.

Syntonic does have a connotation in the field of the physiology of the nervous system and its use is accredited to *Bleuler* who used the term to describe a "stable integrated type of personality," as quoted by *Dorland's* Medical Dictionary. *Syntonic* is also used in the adjective form to indicate a balanced integrated nervous system particularly the autonomic in which division there exists a state of dynamic antagonistic tension between its two systems. When these tensions are equal the nervous system is said to be in syntony or in the syntonic state. It is in the last two senses that the word will be used in the amplification and discussion of the problem being undertaken.

It has long been a matter of common observation that individuals react differently in the same environment and to the same

3

stimulus. Many efforts have been made to account for these differences on the basis of heredity, environment, and even nutrition.
Obviously, no two people except blood brothers and sisters can
have the same heredity, but it is a matter of common observation
that blood brothers and sisters, and even identical twins, do not
always react the same way to the same environmental conditions
nor to the same stimulus.

In all people with the exception of surgically mutilated persons
we have similarity of structure and the same number of organs
and parts which function on approximately the same levels and
in like manner. It would seem, then, that there could be very
little if any difference in response. On second thought, it is a
matter of common knowledge that even machines, which have
been accurately built, do have variations in their ability to exe-
cute the functions for which they were produced. Machines do
not possess any of the many varying processes which are found
in living things, hence variation in execution can only be attributed to imperceptible differences in the structure of their
components.

Anthropologists have spent years making accurate measurements of structural differences in human beings in efforts to
determine, if possible, what part variations of structure play in
the observable differences of response. Enough has been learned
to indicate that to some extent functional differences can be interpreted in terms of recognizable structural differences, yet not
enough work has been done in this field to justify the setting up
of standards that can be considered absolute, because any such
standards must obviously be relative to the accuracy of the
measurement and to the considered judgment of the observer who
undertakes to study function in connection with structural
differences.

Recently a school of thought has come into being which seeks
to account for functional differences on the basis of so-called conditioned reflexes or more properly, *conditioned responses*. This
school developed as a result of the original work by *Pavlov* in
which it was found that stimuli not normally present in an
animal's environment could be used to excite responses identical
with those excited by stimuli that *are* normally present in the
environment. The results of his work indicate that it is pos-

sible in a large number of instances to so "condition" physiological processes that they respond, in a manner similar to reflexes, to the application of what would normally be an abnormal stimulus. The entire behavioristic school of thought, championed by *Watson*, is based upon this "conditioning" concept.

But, does the conditioned response explain recognizable variations in conduct, behavior, or physiologic responses? It seems that it does not because it is a known fact that conditioned responses tend to and do revert to their originally unconditioned status unless the conditioning stimulus is more or less constantly repeated. The result is that a previously conditioned animal or individual will revert to the original unconditioned state if removed from the influence of the conditioning stimulus. If the behavioristic theory is to stand as the sole cause for differences in response it must find some way to explain away this weakness in the theory in view of the facts.

In this research we are primarily concerned with deviations from, and variations in, responses that are not usually considered normal, yet the variations will remain within the physiologic limits of the normal. By this we mean responses that are not due to structural change of cells and tissues in the pathological sense. It must be kept in mind, however, that a long continued functional response, near the extreme limits of the physiologic activity, may result in structural changes which may be considered either as compensatory adaptive changes or perhaps could, in a strict interpretation of the term, be called pathological, because they are not the structural states which would exist in the absence of long continued extreme function.

It is one of the axioms of biological phenomena that "structure governs function." While this is true within reasonable observational limits it is not necessarily a one hundred percent. statement. In the main, we would expect a rabbit to be a herbivore and to defend itself by flight. We would also expect a giraffe to be a herbivore and to also defend itself by flight. Due to structural differences, we would *not* expect the giraffe to be a grazing animal at ground level, nor on the other hand, could we reasonably expect the rabbit to use tree foliage as the source of food. The lion and the seal are both carnivorous animals, but due to structural differences we certainly would not expect the lion to get its food from the sea, nor would we expect the seal to secure its food on land.

The structures of these cited instances differ markedly due to the difference in heredity in the animals. In this sense heredity governs structure. Completing the syllogism we might, therefore, say, "heredity governs function." But, *does* heredity govern all function or functional deviations? Obviously it does in terms of genus and species. But in the same genus and in the same species we find variations of behavior and function. Not only that, but we find wide variations in structure in terms of height, mass of head, size of the chest, relative length of extremities to trunk, and other well known differences.

The "conditioned responses" school of thought maintains that the environment in which the animal is reared accounts for response differences, but few of them hold that environment will cause structural differences. Another breakdown in this school of thought is thus brought to light.

Speaking in terms of animals, a difference should be noted here between instinctive response and the drive or urge leading to that response. These two are not one and the same thing. The urge to commit an act, functional or otherwise, is largely an acquisition resulting from the interplay of two factors, the heredity and the environment in which the animal has been reared. The urge or drive to commit an act can be and is altered by conditions that may exist at the moment or under the influence of which the animal has been reared. But once the urge to commit an act has been strong enough to lead to its execution the behavior from that time on is one of instinct, in that instinct governs and controls the *way* the act is performed. To illustrate these differences, one may have the urge or drive for copulation, but due to existing circumstances or prior environmental training must resist and repress the urge. But if conditions were such as to permit the execution of the drive, then the way the act of copulation is performed is instinctive and is the same for all animals of that genus and species. In other words, one is not governed by his instincts in terms of the will to commit an act, but once the will to commit an act takes control of the individual, its execution takes and follows a very definite form pattern. Instinctive behavior is so much alike in animals of the same genus and species and is so predictable that its study becomes a dull monotony.

It seems, therefore, that environment and its conditioned responses only operate to determine the time and place of the execution of instinctive acts, which latter are inherent largely in the structure of the animal. Does, then, environment alter or prevent instinctive structural manifestations of function, either physiologic or pathologic? The writer's thesis is that it *only* alters the drive or urge and does *not* alter structural instinctive function.

Mendel's researches in heredity have clarified differences of structure. Even within the same family, the members of which live largely in the same home environment is found differences in function. Authorities are virtually agreed that functional departures from the normal within physiologic limits may also be due to neurotic or psychotic states, the which are generally recognized to be a result of early environmental conditions. But, is this the sole factor? Is there a common environmental factor, common to all individuals of the human race, which could and does vary for the several members of the race, or the several members of the same family, which could account for the observed differences?

What are some of these environmental factors which are common, and which ones of them are found to vary as between individuals? The first factor is, obviously, that of heredity which as can clearly be seen, varies as between families, but does not vary in the same family. It seems, then, that heredity cannot be the factor sought. A second factor, the school and family environment should now be considered. Within each family insofar as housing, food, clothing, we find a practical constant for that family. The school environment may be a widely different one for any given individual member of the family when thinking in terms of his home environment. But the school environment is common to all members of the same class. Hence, it would seem that this is not the factor sought. A third factor would be endocrine differences as between individuals. Here we find wide differences between the members of the same family, as well as differences between all individuals of the race. The endocrine factor is inherent in each individual and varies from day to day in that individual, but no criteria have as yet been established which can be used with a fair degree of accuracy for evaluating endocrine differences nor for determining all of the reasons for these differences. Fourth, the food intake does vary as between individuals. It also varies in terms of communities, races, religions, and geographical locations. Food intake also varies with the likes and dislikes of the individual. Here, again, no criteria have or can be set up for the evaluation of this factor.

A fifth factor, which seems to be the most important one, is that of the several forms of radiant energy absorbed by individuals. In one instance this radiant energy takes the form of the well known Hertzian waves, now so commonly in use in wireless telegraphy, and the more common radio transmission of sound. This factor does not vary between individuals because all living things are subject to the absorption of the same energy simultaneously. Another form of radiant energy is the infra-red radiation, a low frequency radiation, whose frequency is too low to stimulate the photoreceptors in the eye. Absorption of infra-red energy results in the production of heat. The absorption of infra-red radiant energy does vary between individuals due to differences in body coloring and rate of metabolism within the individual. But it is not believed that the variation is great enough to have marked significance to the problem under consideration. A third form of radiant energy is the photic range or the so-called visible range. This form of energy varies for each individual at all times, due to selective absorption of light by transmitting or reflecting media or bodies. It seems that here we have a factor which can be studied and should be studied as the one universally variable for all individuals. Other forms of radiant energy are the ultra-violet rays, the Grenz and x-rays, and those of the radium compounds. These radiant energies, with the exception of the ultra-violet, are not common to the environment of all people, although ultra-violet may vary considerably as between geographical locations. Another form of radiant energy, is the so-called cosmic radiation, which is common to, and identical for all persons, hence could have little if any effect as between individuals.

It can readily be seen from the foregoing that the one constantly present, yet constantly variable environmental factor is energy in the photic or visible range of the electromagnetic spectrum. Is it possible to show that variations in this range can and do have effects on functional activity? And if so, *what is the mechanism* whereby functional variations are brought about?

Is it possible by altering or controlling the frequency of the photic range of the electromagnetic spectrum incident into the eye, and as a result of these changes to predict altered function, and to *what extent* and in *what manner?* In other words, do changes in the frequency of energy in the photic range incident into the eye cause altered function, altered behavior, altered physiological responses, and are these variations of a similar kind in all individuals?

This is the problem, a solution of which was sought in undertaking this research.

CHAPTER II

SOME PRIOR APPROACHES

The use of light and its modifications to influence physiologic responses is apparently as old as history, if the records are to be believed. It is perhaps specious to mention that the forty-sixth word in Holy Writ is the word "light."

Genesis—Chapter 1.

1. In the beginning God created the heaven and the earth.

2. And the earth was without form, and void; and darkness was upon the face of the deep. And the spirit of God moved upon the face of the waters.

3. And God said, Let there be *light;* and there was light.

Whereas this mention has no great importance for the purpose in hand it is, however, an indication that light has been considered to be a necessity to life by the human race since the earliest records.

The philosophy of the ancient Egyptians and their religions make frequent reference to the sun as the source of not only light but life, particularly that of plants. There is some question in this regard as to whether the Egyptians recognized the close connection between animal life and its dependence upon plant life. There seems little doubt, however, that the dependence of plant life upon sunlight received early recognition.

The Hindu philosophies abound in references to the use of colored light for the attainment of the contemplative attitude. Insofar as this writer knows there is no particular reference to the use of light or colored light for what is now recognized as a therapeutic purpose.

The first reference this writer has found to the use of modified light for its physiological or functional effect is found in the thirtieth chapter of Genesis from the thirty-first to the forty-third verses, both inclusive:

31. And he said, What shall I give thee? And Jacob said, Thou shalt not give me anything: if thou wilt do this thing for me, I will again feed and keep they flock.

32. I will pass through all thy flock to day, removing from thence all the speckled and spotted cattle, and all the brown

cattle among the sheep, and the spotted and speckled among the goats; and of such shall be my hire.

33. So shall my righteousness answer for me in time to come, when it shall come for my hire before thy face: every one that is not speckled and spotted among the goats, and brown among the sheep, that shall be counted stolen with me.

34. And Laban said, Behold, I would it might be according to thy word.

35. And he removed that day the he goats that were ringstraked and spotted, and all the she goats that were speckled and spotted, and every one that had some white in it, and all the brown among the sheep, and gave them into the hand of his sons.

36. And he set three days' journey betwixt himself and Jacob: and Jacob fed the rest of Laban's flocks.

37. And Jacob took him rods of green poplar, and of the hazel and chestnut tree; and pilled white strakes in them, and made the white appear which was in the rods.

38. And he set the rods which he had pilled *before the flocks in the gutters in the watering troughs when the flocks came to drink, that they should conceive when they came to drink.*

39. And the flocks conceived before the rods, and brought forth cattle ringstraked, speckled, and spotted.

40. And Jacob did separate the lambs, and set the faces of the flocks toward the ringstraked, and all the brown in the flock of Laban; and he put his own flocks by themselves, and put them not unto Laban's cattle.

41. And it came to pass, whensoever the stronger cattle did conceive, that Jacob *laid the rods before the eyes of the cattle in the gutters,* that they might conceive among the rods.

42. But when the cattle were feeble, he put them not in: so the feebler were Laban's, and the stronger Jacob's.

43. And the man increased exceedingly, and had much cattle, and maidservants, and menservants, and camels, and asses.

A close reading of the foregoing quotation conveys the impression that the mere act of placing striped or spotted objects before the eyes of animals would cause a change in the pattern of pigment distribution in the hair.

These changes are now-a-days usually considered to be hereditary, but there are two instances which have been cited to the
writer by *Dr. R. L. Cassell*, of Cadiz, Ohio, who reports that they
were mentioned to him by a veterinarian who observed them.
These instances were that of pure bred stock, horses in this instance, where the color lines and strains were known for a number of generation.   During the act of copulation in both of these
instances, reported to *Dr. Cassell*, the animals involved in the
act visually observed another animal of the same species having
a coloring which was totally unlike the hereditary strain colors
found in the animals involved in copulation.   In both of these
instances the colt carried the color and markings of the observed
animal and *not* the hereditary coloring which would be expected.
These are only two reported instances which might be supportive of the Biblical matter quoted above and are here mentioned purely because they are interesting, and not for their
scientific value.

Hippocrates who lived 400 years B.C. and who is reputed to
have been the father of medicine, is reported to have made use
of sunlight in his practice at that time.   *Herodotus* advised sun
baths for those whose muscles were weak and flabby.   This ad-
vice by *Herodotus* has been repeatedly confirmed in our present
day by numerous investigators.

Pliny, reported editorially,[1] is said to have gazed through an
emerald when his eyes were tired in an effort to "strengthen
them."   There are several other reported instances of the use
of the emerald for this purpose.

Goodman[2] reports that red cloth was placed over windows and
red clothing worn by smallpox patients during the middle ages
in an effort to cure the disease and to prevent pitting.   *Hammer*[3]
in Medical World reports his use of red silk over the faces and
exposed parts of the body of smallpox patients as recently as
1938.   As a result of this use of red silk, he claims quicker cures
and practically no pitting.   The red silk was used in the form of
a mask over the face with holes out for the eyes, so whatever

---

[1] Optical Journal and Review, June 1, 1932.

[2] Goodman, "Basis of Light in Therapy," Medical Lay Press, 1928.

[3] Hammer, "Skin Lesions, Diagnosis and Treatment," Medical World, 37,
2, 107–8.

effect the red light may have had was directly a tissue effect and not an indirect effect influenced by the eyes having red light incident into them. *Marchant* published, in 1493, his brochure, "Kalandar and Compost for Shepherds," in which he mentions the use of various colored lights and attributed 'to them certain curative effects.

In 1777 *Favre* first mentioned the use of sunlight for the cure of ulcers. *Pleasonton*[4] began experiments in 1861, with greenhouses having alternate streaks of white glass and blue glass. He reports that by using this method he increased the growth of grape plants and further reports an increase in quantity and an improvement in quality of the fruit. In 1871 *Pleasonton* was granted a patent, No. 119242, covering his procedure, in which patent he also included the use of violet and indigo glass in addition to the blue, and specified their use with sunlight, and with artificial electric light. Presumptively, the electric light used by *Pleasonton* was the arc, because so far as this writer knows, there was not in use at that time an incandescent source of electric light. It is interesting to note in connection with *Pleasonton's* patent that he found as follows:

I have also discovered, by experiment and practice, special and specific efficacy in the use of this combination of the caloric rays of the sun and the electric blue light in stimulating the glands of the body, the nervous system generally, and the secretive organs of man and animals. It, therefore, becomes an important element in the treatment of diseases, especially such as have become chronic, or result from derangement of the secretive, perspiratory or glandular functions, as it vitalizes and gives renewed activity and force to the vital currents that keep the health unimpaired, or restores them when disordered or deranged.

Notice should be taken that he mentions stimulation of glands in the body, effects on the nervous system, and effects upon the secretive organs of both man and animals as a result of the use of blue light, using either the sun or an electrical source of light.

Pleasonton also reports on the effect of violet light and blue light in the taming of an obstreperous mule which could not be harnessed. This mule was placed in a stall in a barn that was

[4] Pleasonton, "Blue Light and Sunlight," 1877.

otherwise darkened except for the glazing of windows with blue and violet glass. In a few days the mule became sufficiently subdued that he could not only be handled, but could be harnessed and worked.

In his book *Pleasonton* reports a large number of cases of disease in human beings, particularly those accompanied by pain, which were relieved by irradiating the patient and the site of pain by blue light.

A few years after *Pleasonton* published the Second Edition of his book, in 1877, a *Dr. Babbitt*, a physician, published a First Edition of his large book "Principles of Light and Color" 1878, in which *Babbitt* entered at length into the philosophy of color and its use for therapeutic purposes. Not only that, but he made the startling discovery that water placed in bottles of the various colors, red, amber, green, blue and violet glass, had some change made in its properties, such that if an individual drank the water, thus irradiated by setting it in the sun, received benefits similar to those obtained by direct irradiation of the patient with the same colored light. Efforts by the writer a number of years ago to confirm this finding by *Babbitt* revealed the fact that no such responses could be obtained by irradiating distilled water. It was found that minute quantities of organic matter must be present if the irradiated water was to have properties other than those of non-irradiated water. The writer found that mixture of yeast cells or bacillus acidophilous in ordinary tap water, and irradiated by *Babbitt's* method, did seem to show effects when used on rabbits. No carefully controlled experiments were possible on human beings, hence no data were secured.

In 1840 *Ollier* and *Poncet* published a treatise "Treatment of Tubercular Arthritis by Sunlight." Following *Babbitt*, perhaps the first effort to put light treatment on a scientific basis was made by *Finson* between the years 1895 and 1897. During the latter year he published his paper "Treatment of Lupus Vulgarus by Concentrated Chemical Rays of Light." At that time he used a condensing blue lens and focussed the blue light on the tubercular swellings, lupus being tuberculosis of the skin. Later *Finson* learned that the same effect could be produced by using a clear glass lens by omitting the blue color. In the latter instance, however, it was necessary to introduce a water cell in the path of the concentrated rays to absorb the infra-red and heat portion of the spectrum to prevent burns, but the effect on the lupus was not in any way impaired by this filtration. *Finson*, therefore, concluded that the effect was purely one of the absorption of the chemical rays of light by the lesion. At a later date *Finson* found that the rays from an electric arc light, when focussed by a rock crystal lens and a water cell, compressed against the lesion resulted in cures identical with those produced by sunlight. He rightly concluded that the rays responsible for the cure were not only present in the arc just as they were in sunlight, but that they were in sufficient quantities in the arc to be effective.

Between the time of the work of *Finson* to 1918 a number of noted men experimented in this field in efforts to standardize the technique and to determine the type and kind of light necessary. The names of *Russell, Turrell, Sir Leonard Hill,* and *Humphreys* are outstanding names, during this period. In 1918 *Huldschinsky* first reported the cure of rickets by the use of sunlight. Rickets has later been found to be due to faulty calcium and phosphorus metabolism in the body due to a lack of Vitamin D. *Huldschinsky's* work was, therefore, definitely pioneering in this field. It is now known that certain wave-lengths in the ultra-violet light energy, when absorbed by the skin, react on the fats and oils contained therein to produce the necessary Vitamin D for the prevention and cure of rickets.

To catalog all the investigators in the use of light and colored light for therapeutic purposes seems to be unnecessary, because the literature is so voluminous and so easily accessible that any one interested in the historical development may easily locate applicable material. It should be noted, though, that practically all of the references heretofore, except that to *Pliny*, have been in connection with the use of light or color for therapeutic purposes, i.e., the correction, amelioration or elimination of abnormal function or definitely pathological conditions. *Pliny*, as noted above, used the green emerald as a means for "resting the eyes." The literature does not contain many references of like application to the eyes until about 1919. At this time *Barr*, an instructor at Ohio State University, in discussing more accurate methods of refraction of the eyes for the application of lenticular aids, suggested a cobalt fixation spotlight at a relative infinity. The reason he gave for this suggestion being that cobalt glass filters out the central portion of the visible spectrum and transmits only the two extreme ends, the red and blue-violet. *Barr's* contention was, since red light and blue light had a differing refrangibility, that they would be focussed at two different points and that the ciliary body, which controls accommodation, could not focus both of them simultaneously upon the retina. His theory was that this failure would result in focussing the blue light in front of the retina and the red light behind it, if it could have reached such a point, and that the mid-central region would fall somewhere between these two points, presumably on the retina. Of course, in the absence of this mid-central region it is not clear just how a determination of refractive errors was to be made although in practice it did appear that refractive errors determined by static skiascopy, and the cobalt fixation spotlight, were in closer agreement with the subjective finding than was the case when the non-filtered fixation spotlight was used. It is well known that the light adapted eye seems to be more sensitive to the yellow-green region, and the dark adapted eye more sensitive to the blue-green region, both regions being mid-spectral. If *Barr's* theory were true, and if it worked in practice when applied, the spectral region of maximum sensitivity would be the one in sharpest focus on the retina, and presumptively this was the underlying thought behind *Barr's* suggestion.

At a later time in the same year, 1919, *Sheard*, then at Ohio State University, and during the same general course of instruction, suggested that if the light in the skiascope were filtered so as to deliver a yellowish-green light to the retina, that a static skiascopic finding likely would be in very close agreement, not only with the subjective, but also in the spectral region of maximum acuity in the light adapted eye.

As was said above, the writer undertook a clinical application of *Barr's* suggestion and was gratified to note a greater agreement in his findings as was noted above. But when he attempted to *combine* the suggestion by *Sheard* with that of *Barr* there was apparently no rhyme nor reason which could be assigned to account for the erratic findings which resulted. Further discussion of this phase will be taken up at a later place.

In 1929 *Jones* and *Clawson*[5] reported on the effects of filtered light on rats, their findings being in a practical agreement with those made by the writer on rabbits at a prior date. In 1930 *Rogers*[6] reported upon his use of red and violet light for influencing ocular functioning, and quotes *Parker* and *Updegrave* as agreeing with his contentions.

Karrer,[7] of Smithsonian Institution, reported that the growth of plant cells could be altered by exposing the cells to the extremes of the visible spectrum. *Brakeman*[8] reported and demonstrated, before an assembly of the College of Syntonic Optometry, that seedlings invariably leaned toward the blue end of the spectrum and away from the red end. The conclusion reached was that the blue end of the spectrum retarded cell growth, whereas the red end increased that rate of growth, thus producing the distortion.

Pacini[9] states that "the effects of visual stimuli in response to color can be observed upon blood pressure, upon muscular, nervous and mental activity," thus showing that colored light thrown into the eyes does alter systemic functions of the body. He further states, "there are many unfathomed depths in the psycho-physiological phases of light and color and particularly in the therapeutic applications of colored light as stimuli operating through the doorway of vision."

From the foregoing it may easily be seen that the field under discussion hereinafter is not a wholly unexplored field, the writer merely claiming that he was the first to organize the data and to experimentally determine related facts, as they pertain to ocular and systemic responses to light in the visible range, and of the response to various visible light frequencies incident into the eye for the aid of vision, its associated and supportive functions.

[5] Jones and Clawson, Journal of Nutrition, November 1929, pp. 111–153.
[6] Rogers, Optometric Weekly, August 7, 1930, p. 840.
[7] Karrer, Science, 84, 216–8.
[8] Brakeman, College of Syntonic Optometry.
[9] Pacini, "Light and Health," Williams and Wilkins, 1925.

CHAPTER III

THIS APPROACH

Following the suggestion made by *Barr* referred to heretofore, the writer made a search of the literature and found it to be practically silent on the effects of selected light frequency bands incident into the eyes and their effect upon ocular problems. The only information that could be definitely located was that which had to do with what are now known as handicap tests. These tests are those in which a very definite frequency band transmitting filter is placed before one eye with no filter before the other, the purpose of such tests being to determine the stability of positioning of visual axes. Nothing was found at that time which would indicate any other use for selected light frequency in the presence of ocular departures from the normal.

As previously stated, a practical application of the suggestion made by *Barr* apparently worked much better than the former method of fixation upon a letter on the test chart or fixation upon a small white spotlight. It was assumed that the results obtained by using this proposed technique occurred as a result, as has been mentioned heretofore, of the focussing on the retina of the intermediate band between those transmitted by the cobalt filter glass. It should be noted here again that this was merely an assumption, there never having been adduced by the writer any scientific proof that such is the case. Nor is the writer informed of any scientific determination which would indicate it to be other than an assumption.

The application of the suggestion made by *Sheard,* that of using an amber or amber-green light as a source of light for skiascopic examination, caused such erratic findings that accurate determinations seemed to be impossible. These results tended to nullify the assumption previously mentioned, i.e., that since accommodation in the eye could not cause the transmitted red and blue frequency transmitted by cobalt to be focussed simultaneously on the retina, the accommodative function must necessarily be in the state of static rest. Peculiarly enough the frequency band passed into the eye by the skiascope, after its adaptation to meet the suggestions by *Sheard,* is a band which is not transmitted by cobalt and lies intermediate between the bands of cobalt transmission. Here we had, then, on the retina simultaneously the red and blue ends of the spectrum as transmitted by cobalt, probably neither in focus, and the third band in the yellow-greens which may or may not have been in focus on the retina. The erratic determination of refractive errors by this means indicated that it might have been in focus part of the time while it was definitely out of the focus at other times during the same test.

Here was a far different state of affairs then if white light were being used for both parts of the test, that is the fixation spot and the white light of the skiascope, in that there were three definite transmission bands with absorption bands between. In other words, on the retina there now was an interrupted continuous spectrum with an absorption band between the red and the greenish-amber, and an absorption band between the greenish-amber and the blue. As will be mentioned later the *absence* of certain frequencies may have accounted for the phenomena observed.

It should be noted that the condition here is a far different one than that existing in the so-called handicap tests, because these frequency bands were all three present in the same eye simultaneously, whereas in the handicap tests one frequency band was in one eye and the white light continuous spectrum in the other eye, both eyes being open and in a functioning state. The handicap test is at times made use of to determine deviations of the visual axes of the eyes such as those found in phorias, which are tendencies of the eyes to turn, and tropias in which there is a definite objectively determinable turning of the eyes. In the handicap tests we are dealing with extra-ocular functions, extra-ocular here being used to designate functions definitely outside the eyeball. In the conditions of observations, in which the *Barr* and *Sheard* suggestions were being used, we were dealing with a definitely intraocular set of phenomena, hence the two things cannot be interpreted in parallel. It might be well to mention here that, in the early work done, no effort was made by the writer to apply it nor to interpret frequency response in an extra-ocular manner.

These inconstant and erratic findings were found in apparently

normal individuals. It is significant in this connection to note
that the variations and the discrepancies from expected findings
are greater when other methods indicated the existence of some
of the higher refractive errors. Later in the investigation it was
discovered that the tendency to esophoria seemed to be greater
when using a combined red and white handicap test in conjunction with a maddox rod or the maddox double prism, that is, the
esophoria seemed to be higher under these conditions than when
the handicap of ruby glass before one eye was not used. It was
suspected that there might be some irritation caused by the ruby
glass which reflexly increased the tonicity of the medial recti
which would, of course, account for the higher esophoria finding,
if later research proved the assumption.

The first effort to bring order out this chaotic condition was a
search of the literature on physiology. No citations could be
found at that time which would in any manner account for the
results, nor could the information found be interpreted in terms
of the factual findings. Medical literature at that time contained
no references to the effect of light frequency either in the eye or
in the body. Some authorities did, of course, mention the use
of "whole" white light irradiated in the body for therapeutic
purposes. It should be noted that these uses were definitely for
the purpose of correcting or overcoming some abnormal functional or structural condition of the patient.

Among the early efforts to explain the observed inconsistencies
was an approach to it in terms of chromatic abberation. Chromatic abberation is known to exist in all lenticular systems. It is
the phenomenon resulting from efforts to focus light upon a point.
This is found to be impossible because the low frequency light is
less refrangible than high frequency light and results in the
former being focussed at a point more distant from the nodal
point of the lens or lenticular system than is the latter. The
question immediately arose as a result of this well known phenomenon, and the later adding of a frequency not found in the
cobalt transmission, if by using this latter source had an effect
of itself. Or, was the observed effect a purely psychic one? Or
was the effect one of temporal summation of three frequency
bands simultaneously impinging upon the photoreceptors of the
eye under examination? The question posed might have been

solved if we were dealing with inanimate lenticular systems, but in the instances under discussion we were not so dealing. It is true that we were dealing with a compound lenticular system and thick lens optics, but there were the additional and unknown factors of physiological and biological response thereto, before any interpretation could be made either by the patient or by the observer.

Frequency stimulation of living structures was early suspected. A search of the literature revealed the works previously mentioned by *Pleasonton* and *Babbitt*. These works while they are highly philosophical did prove helpful in that they indicated that living structures *did* respond in some manner to selected light frequency bands.

A third possible theoretical solution to the problem arose when thought was directed toward the possibility of faulty energy absorption, or more properly, differing energy absorption of the more extreme frequency bands in the spectrum. It has long been known that the sensation response of differing frequency bands is somewhat different in the photoptic eye from what it is in the scotoptic eye, the former being more sensitive to the yellow-greens and the latter being more sensitive to the blue-greens. Experiments with human beings were crudely undertaken in an effort to determine if the sensation response of the individual, under the two conditions, was really a function of the frequency, or if it was a function of the chemical state of the photoreceptors in the retina due to their adaptation status. Due either to faults of interpretation or experimental errors no significant data were secured in this field.

Finally, a pure assumption was made to the effect that the responses observed were due to the absence of a given frequency rather than to the absorption of the frequency transmitted. Since the function of a light filter is to remove some quality of the light it must of necessity pass other qualities. It is fairly well known that physiological responses can be brought about by removing some qualitative factor from the environment in like or similar manner to that of adding some other qualitative factor. This phase will be discussed later.

Another significant finding was that the degree of discrepancy, quantitatively, seemed to vary in different people under the same type and kind of stimulus. Some biotypes showing more erratic findings than others. Anthropologists long ago proved function to be an attribute of structure and have reduced the proved facts to the anthropological axiom, "structure governs function." Structurally it was found that some were tall and some were short; some had wide heads and some had narrow ones; some had massive barrel-shaped chests, some thin flat chests; some had an excessive ratio of length of long bone to trunk length, while still others had a low ratio of length of long bone to trunk length. Was this difference in human structure the answer to the finding of discrepancies? Efforts to classify people in terms of quantitative as well as qualitative discrepancy at that time availed little, although at a later date the field of anthropology had sufficiently advanced to give a rather definite answer to the question posed above. This will be discussed in Part II hereof.

In an effort to secure sufficient data it seemed necessary to undertake animal experimentation. In conjunction with *Dr. F. F. Wilcox* and *Dr. O. W. Spangler,* experiments were begun with rabbits. Rabbits were kept in the same environmental conditions as to housing in day time and at night, and were fed the same food in like quantity, and were supplied with water of equal quantities and from the same sources. The only variable factor in the various rabbit cages was the light filters placed before the cages. Some startling results soon become apparent.

Among the observed findings were that rabbits under some light conditions lost their fur, some in patches and some became almost bare of hair. Rabbits in other cages under other light frequencies, consistently developed cataracts. Others under other light frequencies developed symptoms which in human beings would be recognized as toxic symptoms, in that body temperature rose and pulse rate increased with every evidence of lassitude with unwillingness to move about. Rabbits in still other cages under other light frequencies failed to gain to the normal weight or to maintain a weight even approximating that of rabbits in other cages. Other rabbits in other cages developed excessive fecal elimination almost to the point in two instances as being classifiable as watery diarrhea. Rabbits under other light frequencies became sterile and failed to reproduce, and the females did not develop oestrus following copulation.

Perhaps the most striking and unsuspected deviation from the normal that was found in that rabbits under some of these frequencies developed excessive length of the long bones, and a weighing of their skeletons after death showed bone weight to have been from twelve to about eighteen percent. higher than that of the bone weight found for the rabbits in the control cages. This finding in terms of length of long bones has also been found to be true in human beings. Anthropologists are agreed that man is taller today than he was a few generations ago. The greatest change in length of long bones having occurred in the last two generations, a period of approximately forty-five years. It is interesting to note in this connection that the extensive use of artificial light from electrical incandescent sources was developed about fifty years ago and has come to be extensively used since that time. Endocrinologists are in virtual agreement that the length of the long bones and the age at which the epiphysial plates unite with the shaft is controlled by hormone secreted by the anterior pituitary. Was the anterior pituitary caused excessively to secrete this hormone necessary to cause this excessive growths of long bones? In other words did the light frequency used influence the pituitary?

Referring back to the changes observed the reader should be reminded that most of these changes required rather long periods of time for their production, times varying from three months to eighteen months. The degenerative changes mentioned above usually appeared rather early and obviously any development having to do with the growth of long bone had to be one which took place early in the life of the rabbits under investigation.

Again the question posed itself. Did these effects result from the frequency transmitted into the cage or did they result from the absence of the frequency absorbed by the filter? That is, was it the absence of some frequency which resulted in the changes, or was it due to stimulation or inhibition of function by absorption of the transmitted frequency? It must here be admitted that to date there has been no adequate answer found to this question. It is known, however, that there is just as marked physiological reactions to the absence of some environmental factors as are found when new environmental factors are added. This point may easily be demonstrated in the following manner:

If an individual is stood on the floor with his feet separated about one foot, and asked to assume a position of static equilibrium, and then has someone apply pressure to his shoulders in an effort to push him off balance, the subject having been previously instructed to maintain his vertical position against the opposing force, he will be placed under increased tension, yet remain statically vertical. If he succeeds in so doing for a period of time, and the position is maintained after the applied stress, it will be found that a sudden removal of the impressed force to the shoulder will cause a violent reaction in the direction from which the opposing pressure was applied. This reaction may be so violent as to cause the individual under test to completely lose his equilibrium and he may fall unless supported. This experiment demonstrates the instant and violent reaction to the sudden removal of an environmental factor to which the individual had previously adapted himself.

In an effort to clarify the writer's thinking in this matter the question of absence of frequency removed by the filter and its possible cause for the response was posed to *Dr. William Beck,* Department of Biology, University of Dayton. *Dr. Beck* expressed it as an opinion that the responses noted were due to the physiologic compensation by the body for the loss of the stimulation previously experienced by the radiant energy which was now removed by the filter. Some experiments were undertaken to see if possibly experimental evidence could be adduced in proof of this point. Goldfish were kept in aquaria under red light for long periods of time. Red light should be thought of in this connection as the absence of blue, green, and amber light, that is, a minus blue, green, and amber. Fish in these aquaria all developed a fungus growth on and between their scales. Removing the fish from the aquaria and immersing them in mercurochrome in an effort to destroy the fungus, following which immersion they were placed in the filtered light aquarium, proved unavailing as a means of removing the fungus growth. Furthermore, these fish all developed cataract and all died. It could, therefore, be said that visible light that is minus blue, green, and amber, permits growth of fungus on fish between their scales, and results in ocular cataract in fish and other degenerative conditions. Unfortunately, it can also be said that fish kept under red light undergo the same changes, so it seems that to date it is still impossible to give a definite answer to the question as to whether the reaction results from the absence of a frequency or its transmission.

Rinck, at the suggestion of the writer, repeated the foregoing experiments using an unsaturated red, i.e., a pinkish tinted glass for the aquarium. Even in the presence of an unsaturated red, the fish developed fungus, the scales lifted away from the body and all the fish died in approximately ninety days.

It cannot, therefore, be successfully questioned that there is some effect upon living structures as a result of the absorption of certain selected light frequencies, or because of a reaction to the absence of certain bands.

Also, as a result of our experiments with the rabbits it cannot successfully be denied that it is not necessary that the tissue or eyes themselves be irradiated, it being sufficient, in many instances, to merely irradiate the drinking water given the animal. This series of experiments was undertaken as a result of *Babbitt's* discoveries that as a result of water placed in colored bottles and then irradiated by the sun, this water, so irradiated, later being given to patients resulted in therapeutic effects. In the writer's experience like effects could be produced in the rabbits by irradiating ordinary tap water or water taken from the river, but could not be produced by irradiating distilled water. In this respect his findings were in agreement with those previously reported by *Babbitt.* Naturally the question arises as to what happened to the water as a result of irradiation? In the writer's opinion nothing happened to the water because no effects were produced by the administration of irradiated distilled water, but effects became quickly apparent when water containing organic matter was irradiated. The conclusion is that the various light frequencies used produce some change in the *organic* matter either in solution or in suspension in the aqueous medium. An interesting point for consideration here is this: If the changes which take place *are* in the organic matter in the water, do these changed organic compounds have the power of transmission of the light frequency effect to other living organisms? Obviously, experimental findings indicate that they do. But experimental findings do not indicate the mechanism whereby these effects are mediated.

Obviously, then, the problem before us involves organic matter, the unit being the cells themselves, primarily, and perhaps secondarily involves some nervous response as a result of direct stimulation of nerve receptors in the eye itself. A study of cellular phenomena seemed imperative and was undertaken in view of this concept. Some evidence in support of the theories was found, but cellular changes and cellular modified function as recorded in the literature could not and do not account for all of the observed responses. Since in higher animals most responses are mediated through the nervous system it seemed necessary to undertake later some study of neurology in this connection.

In Part II hereof will be found a review of the essential facts of psychology, neurology, endocrinology, cell and tissue response, individually and collectively, as they alter and affect the systemic reactions.

PART II

CHAPTER IV

CYTOLOGY

Living things are characterized by certain attributes which differentiate them from inanimate ones. Life perhaps does not exist as an entity despite efforts of philosophers to so classify it.

The two fundamental theories of the origin of life other than the religious or philosophical tenets are classified into two groups. The first group might be termed the vitalistic group, in that this group's theories hold that life only comes into existence as a result of prior life, or is transmitted by some prior living thing. One of these theories was enunciated by *Beck* and is to the effect that life on this planet may have been brought here from some external source in the form of a spore, which, finding a proper environment for its development, grew into some form of fungus life. Due to accepted theories of evolutionary changes, this parental living spore evolved into the multiplicity of forms of life as we now know it. It is not the purpose of this thesis to defend these theories in terms of religious or philosophical thought, it seeming merely necessary to mention that it might be difficult to separate complete thinking along this line from the Biblical statement of origin. A close reading of the first few chapters of Genesis states the order of origin of the species in a virtually identical manner with that conceived and developed by the so-called evolutionists. Hence, as this writer sees it there is no direct conflict between *Beck's* stated concept of origin and the commonly accepted religious statement.

The other group of theories as to origin may be called the mechanistic or the physico-chemical group. *Von Bayer* in about 1828 reasoned that in the original environment on the earth there must have existed water vapor and carbon dioxide gas. *Von Bayer* reasoned that if this vapor and gas could be caused chemically to combine, the combination would be some one of the well recognized organic compounds, all of which contained carbon, hydrogen, and oxygen. *Von Bayer's* reasoning caused him to see that the first step in this organization would be the production of formalin, which if kept under the same conditions would conceivably break down into one or more of the invert sugars. *Von*

Bayer made diligent effort to produce this reaction, but due to the then lack of knowledge of what is now known as synthetic chemistry he failed to cause the reactions he foresaw.   At a later date, one of *Von Bayer's* students succeeded in producing the chain of reactions conceived by *Von Bayer* by making use of a uranium salt, and by placing a mixture of carbon dioxide in water and the salt in sunlight.   Due to the absorption of the energy of the sunlight by the mixture of water and carbon dioxide in the presence of the catalyst, the uranium salt, he succeeded in doing what his preceptor had failed to do.   It is true that the production of an organic compound from two inorganic compounds is not the production of life, it merely being a physiochemical process for producing one of the end products usually produced by plants.

In about 1900 *John Uri Lloyd* succeeded in confirming the work of the student of *Von Bayer* and went much further into this field of the synthetic production of organic compounds. *Lloyd's* study led him into a field now recognized as colloidal chemistry for which work *Lloyd* is more or less universally accepted as the father of modern colloidal chemistry.

It is apparent that both of these theories of the origin of life had reliance in the existence of a source of light energy.

The higher animals used to be considered to be unitary by early students until in the latter part of the 19th century, *Schwann* enunciated his cell theory as it applies to the higher animals. *Schwann's* contention was that the higher animals were composed of minute living entities which he called the cells.   At that time scientists were prone to ridicule *Schwann's* theory that man could be composed of a multiplicity of individual living units.   Later research, however, soon confirmed *Schwann's* contention and the biological world today accepts the living cell as the unitary component of all of the complex living organisms.

Briefly, the cell consists of a semipermeable boundary, the cell wall.   Within this wall is found a mass of matter known as protoplasm which possesses certain powers, when organized within its wall, which are necessary for its continued life, growth, and the reproduction of its kind.   The cell also contains in its protoplasm a somewhat more dense mass known as the nucleus which is apparently the control center for cell activity.

This cell unit, due to the semipermeable character of its wall, has the power to absorb from an aqueous solution such elements and compounds as may be necessary for its life processes. Not only that, but it also possesses the power of passing from its interior into its aqueous environment such products of its own activity as may have been produced by its life processes, which, if retained, would be detrimental to a continuation of its life.

Living cells as units possess one characteristic not found in dead cells nor in inanimate matter. This characteristic is the ability of the cell to make changes of form, structure, shape, or position as a result of changes in its environment. This attribute is called irritability, and is the distinguishing attribute between a living and a dead cell.

These cells possess another quality which is important for their continued life and that is their ability to repair themselves and to maintain their cellular integrity. Illustrative of this ability is the power of the cell to reform those portions of its cell wall which may have been accidently broken or experimentally destroyed. The exposed protoplasm, at the point where the cell wall was broken, possesses the ability to reform a new cell wall, thus enclosing the living protoplasm within its proper semipermeable wall. This power is possessed by the cell even if it should be torn into two complete parts. Each part of such a torn cell will reform the semipermeable boundary wall.

The only requirements for the continued life of the cell is that it be in an aqueous solution containing its necessary food and that its waste products, which are passed out through the cell wall, shall be removed from the immediate vicinity of the cell. When these two conditions are properly maintained, assuming a proper temperature, the cell will continue to live, to grow, and eventually will reproduce its kind. Theoretically cells live indefinitely. Perhaps the greatest experimentally controlled life of a mass of cells under the conditions mentioned is the well publicized piece of chicken's heart of *Carrel's* experiment. This bit of chicken's heart was excised on the 17th day of January, 1912, and still continues to live and reproduce its kind in the laboratory, merely requiring that proper temperature be maintained and that it be supplied with food and that its waste products be periodically rejected from the aqueous solution in which it lives. A mathematically inclined scientist has calculated that if all the cells which were reproduced by this bit of chicken's heart, and if none of them had been rejected by the laboratory technicians, that on the date it was twenty-five years old there would have been enough cells to have covered the earth to a depth of one inch. This calculation gives some idea of the rapidity with which these cells reproduce their kind, for obviously these cells are very small and can only be seen microscopically.

The ability of the cell to live in an aqueous solution depends upon the well known phenomenon of osmosis. This being the phenomenon of passage of solutions through membranes. Strictly speaking, the semipermeable cell wall is not a membrane, because, histologically, membranes are composed of cells of like kind and functional characteristics. It is, however, easier to think of this semipermeable boundary of the cell as a membrane when thinking in terms of the phenomenon of osmosis. It should be mentioned here that the phrase "aqueous solution" has been used advisedly above, because only elements or compounds in solution may undergo the phenomenon of osmosis, with perhaps the sole exception of what are known as colloidal solutions in which the size of the particles of elements or compounds have been mechanically reduced to an approximation of molecular size. The phenomenon of osmosis may be to some extent varied by externally applied pressure or by the existence of differences of electrical potential, both of which alterations are well known.

Animal cells, that is cells found in what are ordinarily recognized as animals, while living, possess the property of accepting potassium compounds at a much greater proportional rate than they accept the sodium compounds. In the living cell this ratio is about 60 : 1, potassium to sodium. The cells in the human body contain chloride of potassium in their interior, but they live in an aqueous environment containing chloride of sodium. So long as this status exists the cells function adequately. In the presence of any condition which causes a cell to permit entrance of an excess of chloride of sodium to its interior there results eventual death of the cell.

One of the phenomena of living cell activity is the perpetual rate of its absorption of food and its expulsion of waste. In dead or inanimate states, a state of equilibrium is soon reached between solutions within the cell wall and those exterior to it. The president of the Royal Society of Science, *Sir Frederic Gowland Hopkins*, in his presidential address before the society on September 14, 1933, made this statement: ''Were it not equipped with catalysts every living unit would become a static system.'' Catalysts are those elements or compounds which are necessary to be present to enable a chemical reaction to take place, but which themselves do not enter into the reaction. That is, the catalyst remains in its original form and state after the reaction takes place just as it was prior to the reaction. Obviously, under these conditions, some catalyst is a permanent part of the living cell and possesses a sort of ''trigger action'' which repeatedly and perpetually results in the phenomena exhibited by the living cell. Since the chloride of potassium is a constant finding in animal cells some biologists are inclined to the belief that it is the potassium ion which serves as the permanently contained catalyst.

An animal cell may have its activity increased by a number of environmental conditions:

A. An increase in its oxygen supply. Oxygen is necessary for the maintenance of life of the animal cell because by its combination with carbon and hydrogen compounds, it produces heat and energy, heat being essential within certain limits for the maintenance of the life of animal cells. A distinction here should be made between animal respiration and cellular respiration. Animal respiration may be defined as the making use of special organs within the animal body for the absorption into some aqueous medium of atmospheric oxygen. Cellular respiration is the acceptance from this aqueous medium, of the oxygen it contains, by the cell, following which intra-cellular oxidation takes place with the eventual elimination from the cell of carbon dioxide as one of the end products of this combination.

B. Food in solution will also increase the activity of the cell by supplying it with elements for its own growth, as well as for the production of energy. Food for the animal cell must, among other things, contain nitrogen in compounds; the presence of the nitrogenous compounds being one of the distinguishing marks between animal and most plant cells. The proteins used as food are nitrogenous compounds, but in their pure state, as usually taken into the stomach cannot be utilized and must chemically be broken down into more simple compounds, among others, the amino-acids, before they can be utilized by the cell. So simple a food as raw white of egg may be definitely poisonous if it gets directly into the blood stream without having previously undergone the fractioning process of digestion. *Eccles*[1] has shown that so small an amount as one ten millionth of a gram of raw egg white is fatal to a previously sensitized guinea pig.

C. An increase in rate of waste removal from the immediate vicinity of the cell increases its rate of living because all cells die in the product of their own activity unless those products are rather promptly and completely removed from the cell's environment. In the higher animals this removal is a function of the lymph systems and the blood stream.

D. The presence of a certain amount of heat is also necessary to maintain life in animal cells. In the human being this optimum temperature ranges somewhere between ninety-eight and one hundred degrees, Fahrenheit. Temperatures beyond either of these limits result in functional and rather severe systemic involvement.

E. Externally applied pressure may increase the rate of absorption of food or the elimination of waste, thus increasing activity. Liquids are not compressible in the ordinary sense, but it is easily demonstrated that the application of an externally applied pressure to one side of a semipermeable membrane very quickly will result in an increased osmotic rate, which can be made temporarily to reverse the normal direction of osmotic interchange. Of course, this is only a temporary state and will return to the regular order of osmotic travel in the presence of long continued pressure.

F. The acid-base status of the aqueous environment of the cell may be altered so as to increase cell activity. In the living human body the optimum acid-base status is in the region of pH 7.4, which is slightly alkaline. If the alkalinity

[1] Eccles, Eclectic Medical Journal, September 1933.

should increase to a point of about pH 7.6 there is a definite lessening of the systemic powers to respond to external conditions. Also, a pH of this amount predisposes to the allergic diseases. Such a high pH is technically known as alkalosis. *Contra,* an acid-base status as low as pH 7.2 is definitely less alkaline than normal and is known as acidosis. This state interferes with cellular metabolism and causes systemic responses indicated by sypmtoms which are easily recognized by physicians.

G. Any change in the local electrical charge may alter cell activity. The living cell carries in its interior a negative charge relatively to the charge carried by its aqueous solution. The nucleus within the cell presumptively carries a positive charge in reference to its own negative protoplasm. Any condition which alters the degree of potential difference may lessen or increase cellular activity, i.e., lessening the relative negative charge within the cell slows its activity, while increasing the relative negative charge augments cell activity. This latter condition is believed to be one of the factors responsible for cancer.

During an increase in cell activity due to any one of or to a combination of the foregoing, there co-exists an increased permeability of the cell wall. Clearly, this is necessary because if the cell must do more work it must also ingest more food, and if it ingests more food and does more work it will produce more waste products to be eliminated from its interior.

Cells may be caused to do more work aside from the foregoing by the application of a stimulus. A stimulus has been defined by *Thompson,* of Magill University, as follows: "Any change in environment which produces an active reaction in the cell is called a stimulus." Attention is directed to the first word "any," and to the phrase "active reaction." It makes no difference what the change in the environment may be, if the cell actively reacts to that change, the change constitutes a stimulus.

Each cell is irritable to a stimulus after its kind. A muscle cell only contracts as a result of a stimulus. A gland cell secretes when stimulated. A nerve cell may initiate an impulse or modify the impulse as a result of the stimulus received by it. Each of the foregoing cellular responses is identical for a given cell regardless of the type or kind of stimulus. Stimuli are of many kinds. They may take the form of heat, radiant energy, solutions, mechanical contact, chemicals, cold, electrical or any other form. Summarizing, it may be said that there are many stimuli, but for any given cell there can only be one response regardless of the stimulus used.

During reproduction by cells they, of course, reproduce after their kind, so it may logically be said that the kind of response by a given organism or cell is fundamentally a function of its hereditary origin regardless of the environment in which it is placed.

The biological law governing the situation has been described as the biological triangle, the base of which is the *heredity,* one side of which is the *environment,* and the third side the *response to* environment, as determined by the preceding two. Not only does the biological triangle apply to individual cells, but it applies to the complex organisms. Since heredity governs structure, and structure governs function, we find in human beings that variations in structure do result in different types and kinds of responses to the same stimulus. Obviously, the study of the heredity of an individual human being is almost an impossibility, because very few families have genealogical records which are sufficiently complete to set forth the individual reaction characteristics of the listed members of the family. The only clue to an anticipation of possible reactions by a given human being is the objectively observable structure. Types of structures are classifiable. These structural classes have been given the name "biotypes," i.e., the biological type of the individual under observation. For this reason it is necessary to have a fairly clear picture of what constitutes the various classes of biotypes and the type and kind of responses to be expected in each of the biotypes. Criteria for these determinations will be set up hereinafter.

All living cells are bound by certain reaction characteristics in terms of function, structure, the influence of function upon structure, and the influence of structure upon function. The first of these characteristics may be summarized by the following statement: Normal cell structure is essential to normal function. Amplification of this statement must embody the thought of the survival of the fittest, strongest, most healthy, most productive, and perhaps the most important, the most *adaptive* to environmental conditions, the biological law here being, "Adapt or die." *Gerassimow* conducted his experiments with single cells in efforts to demonstrate the above principles. Separating a cell into two parts, one of which contained the nucleus, and the other having no portion of it, resulted in two living structural cells with widely different characteristics. The cell without a nucleus grew at a rate of about one-twentieth to one-two hundredth of the rate of the cell containing the nucleus. The enucleated cell lost its ability to solutionize starch. The semipermeable boundary of the enucleated cell was less extensible than in the cell containing the nucleus, and finally the enucleated cell remained pale in color and obviously could not reproduce its kind due to the absence of a nucleus. This experiment by *Gerassimow* demonstrates the fact that in the presence of normal cell structure we have the only possible chance for normally functioning cells.

A second reaction characteristic of the cell may be stated as follows: Continued abnormal function results in abnormal structure. Cells forced to function outside their normal physiologic functional ability sooner or later alter their structure, that is adapt themselves structurally, to meet the new conditions. For instance, a habitually exercised muscle does increase in size, and in the number of its cells, with a resultant increase in power of the muscle. However, forcing this same muscle to continue in operation at or near its physiologic limit results in the formation of connective tissue or fibrosis, if it be a striped muscle, and results in a comparable condition known as sclerosis if it be plain muscle.

A third characteristic may be stated: A normal cell environment is essential to normal function. Obviously, the weeding out of the unfits by unfriendly environmental conditions results in a remaining sturdier stock which is stronger, more highly vital, and more adaptable to environment. This characteristic, obviously, would be suspected if one thought in terms of the two precedingly discussed characteristics.

Another characteristic, the fourth, is the principle of cell selectively, also called the habit of the normal. This principle may roughly be stated in this manner: A cell or a tissue given a choice between two acts always does the one requiring the least expenditure of its own energy.   This law goes even further in that it applies to the selection of food from the environment when more than one occurs therein.   In the latter instance, the cell selects that food from which it can obtain the most energy with the least expenditure of its own.   An example of this is the utilization by the cell of alcohol when sugar is also present in its food supply. The body as a whole obeys the same law as does the cell in terms of selectivity.   Due to the fact that every effort is made by the cell to retain its high energy content, with the least possible expenditure of its own, we find that cells, tissues, organs, and in animals having a nervous system that habits result, they being merely a lessened liminal value of the nerve cells and reflex paths. This liminal value having been so lowered that what was previously a subliminal stimulus now becomes liminal and elicits a response.   Another interesting and important outgrowth of this principle is that the greater the enforced departure from the normal, as it pertains to function, the greater is the tendency for the cell or tissue to revert to its normal status.   It seems that here we are dealing with what might well be thought of as a dynamic antagonism between the cell in an abnormally functioning state, and the force or forces which caused it to function abnormally.   It is this phenomenon which aids in the restoration of health to an ill body.

Another one of these characteristics may be stated as this: Nothing of value can be added to the normal environment of the cell.   Another way this might be said is that once the normal environment has been established for a given cell, the addition of another environmental factor actually lessens the power of the cell to live and function.   Illustrative of the foregoing, paramecia may be cited as a characteristic instance.   In order to raise paramecia in the laboratory requires water, light, proper temperature, certain minerals in solution in the water, and bacteria or some nitrogenous compound in suspension in the water.   Under these conditions paramecia will live, grow, and reproduce.   Anything other than the foregoing which may be added to the environment requires some adaptive process on the part of the organisms to meet the change, failure to do which results in death.   Geologists have found a striking instance of form change, as a result of the addition of a compound to an aqueous environment, in the strata of rock above the Great Salt Lake. In strata at the higher elevations are found fossilized algae of the typical fresh water type as they now exist. At a lower elevation is found a type of algae whose structure indicates that it was in a transitional state between a true fresh water type and a true salt water type. At still lower elevations, the fossils are of true salt water type algae. Still lower in these strata the number of fossilized algae show a marked decrease, while the Great Salt Lake itself today contains no living algae. Here we have a picture of a gradual addition of salt to the water of the lake. The algae adapted themselves to the changing salt environment as long as possible, but eventually a point of salt saturation was reached which was so high that no adaptive change could be made, which resulted in the death of all of the contained algae. *Wistchi*, of the University of Iowa, hatched frogs' eggs under different temperatures and found that at sixty degrees, Fahrenheit, the ratio was eighty-four males to each one hundred females, but when temperature of the aquarium was raised to eighty degrees, Fahrenheit, the ratio was two hundred males to *no* females. It appears, then, that so simple a thing as a temperature increase of a mere twenty degrees could be responsible for complete annihilation of frogs in our environment, due to an absence of females.

During adaptation to a changing environment cells may undergo a form change, or an alteration of the type and kind of food ingested, or a change in the composition of waste products eliminated. Obviously, any one or all of these conditions might take place. A restoration of the originally normal environment tends to result in a return to the original forms, food uses and character of waste eliminated. The human body as a whole tends to obey this same law of adaptation, the body being made up of interdependent parts, each of which obeys the above law.

A sixth principle is stated in this manner: Nothing better than the normal environment can be provided for injured or abnormally functioning cells. A practical instance of this is in the case of nerve degeneration following anterior poliomyelitis. Here the normal environment of the motor nerve cell, the motor fiber, and the muscle is a response by the muscle to the receipt of an efferent impulse by its motor cell. In infantile paralysis the nerve cells have been injured by the disease process, but it has been found that if the muscle, for which it is a motor cell, is exercised by passive massage, by electrical stimulation or by voluntary motion after removing the pull of gravity by immersing the member in warm water, it is possible to prevent complete degeneration of its motor cell and, in a large number of instances, to restore in part its functioning power.   In other words, the maintenance of the normal proprioceptor environment to the motor cell, by causing the paralyzed muscle to function, results in actually aiding the injured cell to recover.

Jacques Loeb succeeded in growing hydranths in daylight, but not in darkness.   He also found that he could grow them under blue light but not under red light.   Obviously, a hydranth kept under red light, and which had begun to undergo degeneration, could be restored to approximation of normal by placing it under blue light.   Incidentally, in connection with the use of red and blue light it has been demonstrated by the writer and others that the rate of germination of seeds can be altered by prior irradiation with either of the extreme ends of the spectrum, and that subsequent irradiation by the opposite ends of the spectrum cause a reversion to the normal rate of germination.

Another characteristic of a living cell is that, after its reserves are exhausted, continued activity calling for an expenditure of energy takes place at the expense of the cell structure itself.   The cell utilizes its own protoplasm as a source of energy, resulting in eventual atrophy, if the demand be continued over a long period of time, and if no other food supply is established.

An eighth characteristic is stated in this way: Abnormal function indicates either abnormal structure or an abnormal environment, usually the latter.   Enough has been said above to make this statement clear without further elaboration.

A ninth characteristic which applies to the more highly organized organisms is that symptoms exhibited during departures from the normal are merely evidence of efforts to continue to live under these conditions.   That is to say that the symptom is merely an exhibit of functional efforts to overcome the abnormal condition.   In this sense, disease is not an entity, for diseases are only recognized by the symptoms, which in turn should direct the attention of the physician to the *type of effort* being made by the body.   A correct interpretation of this effort is the prime requirement for adequate therapeusis and should indicate its own remedy to a properly trained clinician.

Chapter V

NEUROLOGY

The study of neurology has proceeded along a number of lines. Some of these approaches have been purely on a gross anatomical basis, others upon a histological study, particularly of degenerated fibers, in an effort to determine communication paths within the nervous system. Still other studies have been made by the ablation of certain portions of the brain or portions of the cord in an effort to locate the site of the localized function control. All of the preceding methods have proved of value, yet none has proved wholly complete in so far as the neurology of man and his functions are concerned. In more recent years an approach to the nervous system of man has been made in cases of disease of the nervous system. In these cases complete histories and syndromes have been kept and developed with the thought in mind of eventually examining the brain and cord structures *post mortem,* if consent for such examinations could be secured from relatives. This study has resulted in a more or less profound knowledge of how the brain and nervous system functions, particularly in a negative manner, however, that is by a study of the loss or failure of function.

Roughly speaking, in the mammals we have two nervous systems which interlock, interact, and are mutually supportive, despite the fact that these two systems have distinctly different classes of functions to perform. One of these divisions of the nervous system is the cerebrospinal system, consisting of the major portion of the brain, cerebellum, and spinal cord. This division is so set up that it is quickly reactive to stimuli, whether the stimuli be received over one or more sensory paths or not. This system is the extero-fective system in that it enables the animal to quickly meet changing conditions in his environment by muscular activity or other activity under conscious control. The effector of this system is striped muscle, which has a high speed rate of response to the receipt of a nervous stimulus.

The other major division of the nervous system, the autonomic, is designed for rather slow responses, and, as will be shown later, consists of two dynamically antagonistic divisions known as the sympathetic and parasympathetic. Of these two divisions the parasympathetic is much the more specialized, but less continuous in its effective responses. This latter division arises largely in the midbrain in the vicinity of the vasomotor, cardiac, and respiratory centers, which centers are not sharply defined, yet function specifically from the physiological standpoint. The autonomic is the intero-fective system in the mammal.

Theoretically, the nervous system consists of units known as neuromeres, which schematically consists of some form of a receptor, an ingoing conducting fiber, a nerve cell center, an outgoing conducting nerve fiber, and something to respond to the receipt of the nervous impulse. This neuromere, the unit, is merely the well known "reflex arc," and is capable of functioning directly as such, but in order for an outgoing impulse to be set up there must have been a prior incoming impulse. Later we shall show that it is possible for the nerve cells in the neuromere to function automatically in the absence of an afferent impulse stimulus, but for our present purpose, this fact will be ignored. We shall, however, consider the neuromere as the theoretic unit of the nervous system.

It must be remembered that the entire nervous system is composed of millions of those units each having its own specific function to perform, and, in almost all instances, each neuromere is in turn connected to other neuromeres so that to some extent they are all interacting. This latter interacting process constitutes the act of integration or unifying response. This act of integration is the all important function of the nervous system, failure of which function is some departure from the normal in some one or more of the physiologic functions.

The mammalian central and peripheral nervous system has been compared to a mammoth telephone exchange system, consisting of many receiving wires, and many outgoing conducting paths, with a switch board and the operators, or the mechanical circuit selectors which determine just where an incoming message shall end. These incoming messages proceed from a receptor, activated by the human voice, which travel down a wire, the afferent nerve path, pass into the switch board, corresponding to the nerve cell and its dendrites, and out over another wire, the efferent path, to a telephone receiver corresponding to the effector, and which determines the type and kind of final response. Obviously, the connections may be very complex, in fact they are most complex in the normal body, because of the multiplicity of actions and reactions which must be integrated into a unified activity by the whole organism. Every nerve cell in the body and its dendrites and axones operate for the carrying on of these reflex activities.

Like all analogies the telephone exchange analog breaks down when we attempt to carry it over into physiological functions, particularly in view of the smoothness and complexity of operation of the mammalian nervous system.

In the lower levels of life we see rather simple and limited movements which are effected by a rather crude nervous network. In the jellyfish these movements are slow and rhythmic and have an order of amplitude and rate comparable to that of smooth muscle in the intestinal tract of man. Moving higher in the scale of living things where more rapid adaptation to environmental conditions is necessary, we find concentration of cells in the ventral region with some evidence of segmentation or division into response groups within the animal.

In the higher animals, where rapid movement and postural adjustments become necessary, we have a highly complicated and involved set of reflex motor functions. These postural reflexes are usually integrated and correlated in the cerebellum. The cerebellum is a constant feature of the vertebrate brain and is found as low down in the order as the fishes, birds, sharks, and reptiles. The crossed pyramidal tracts are a later development in the mammals.

As a result of this nervous arrangement, calculated to produce speedy movements, we find that in man we may walk, or run, or engage in extreme activity such as some of the athletic events requiring an extreme of coordination within an extremely short period of time. On the other hand, we may find a very high degree of coordination and correlation of activity in the same individual while he is playing some musical instrument, where the movements may be very slow, but where coordination and a sense of timing is as important, or even more important, than in the more violent forms of movement.

While the muscular movements mentioned above may appear

to be complicated and require a nicety of integration and correlation, these are nothing when compared with the complex associations and integrations in the cerebral cortex, to say nothing of the *integrations* in the basal ganglia and in the mesencephalon.

Integration of action is the all important function of the nervous system. At the lower levels in the body it is a purely automatic and relatively fixed procedure, such as the reflexes in the posterior root ganglia in the cord. At the higher levels the nerve centers apparently have the power to interpret impulses in terms of past experiences, which result in certain individualized activities or reactions. Also, in the higher levels is a certain individualized method of learning which may perhaps constitute the intellect of the individual *per se*. At the lower levels the response, while relatively fixed, may be modified by certain variables within the system, but any such modification never takes a form comparable to interpretation of impulses at the higher levels as indicated above.

Once a nerve center receives an incoming impulse over an afferent path, the response which it sends out over an efferent path expresses the reaction in terms of variables inherent within the nervous system, as well as higher center interpretation variations. The total of these variations from pure reflex responses constitutes the behavior of the individual.

Now, if we can imagine the multiplication of these neuromeres and the intero-fective and conditioned alterations of the pure reflexes, then multiply this thousands of times, we would begin to get some idea of how the brain works and the complexity of the nervous responses in any one individual. This is, of course, an incomplete and perhaps an inaccurate picture of the complex forms of behavior, because, after all, the cord itself, aside from any connection with the brain, is so anatomically constructed as to make possible some of the rather complex forms of behavior. In reality, the simpler reflexes appear later during the life of the animal and are not developmentally present, since the simple reflexes are influenced to a great extent by the larger behavior patterns. Even so, the neuromere, i.e., reflex arc, is the best basis for attempting to understand the nervous system and should be considered a small working model of the whole.

In addition to the cerebrospinal system under discussion, there is also the autonomic division of the nervous system, which is much more primitive in its responses, although the same reflex mechanism exists in the autonomic. Also, it has these two big differences, however, that it is not so susceptible to having its responses altered by "conditioning," and is not at all affected by the act of cerebration or bringing the intelligence to bear upon its functions. In fact it cannot be altered by the latter. The word "autonomic" implies that it is self-governing, and we find in the mammal it to be just that.

Both the cerebrospinal and the autonomic have a segmental arrangement in that cells at certain levels affect tissues in practically the same level. In a quadruped, walking on all fours, this segmental arrangement takes virtually the form of vertical planes passed through the spine at the level of each vertebra. Between each of these planes we find a segment of the cord with two pairs of root nerves, i.e., two anterior motor nerves, and two posterior sensory nerves, the latter with their root ganglia. These two roots on each side join to form the spinal nerve which is a mixed nerve containing both motor and sensory fibers and is distributed to the regions bounded by the planes superior and inferior to the single vertebra and the circumference of the body between these planes. This segmental division possesses automaticity in each segment, in that stimulation of a sensory nerve to a given segment results in the release of a motor impulse from the root ganglion of the same segment, or from the cells of the cord with which it is connected. If the sensory stimulus is great enough, impulses may travel superiorly in the cord to the brain and will result in some conscious response to the stimulus.

This spinal cord segmentation extends from the base of the skull down through the sacral to the coccygeal nerve, and, as has been shown is relatively easy to visualize in the quadruped. In a biped like man, the extremities, while they hang perpendicularly and parallel to the body, are likely to be in the same vertical plane with the cord. This makes it somewhat more difficult to visualize the segmental distribution. The problem is easily simplified and visualized by mentally placing man in the position of the quadruped.

The segmental reflexes of the cord are normally and naturally present, but some of them are such that they cannot be studied until after transection of the cord above the segment under study, that is, between it and the brain. Such a transection isolates the portion of the cord under study from any connection with the higher centers. Such a transection results in purely cord segment exhibits of response since it is left to function alone and uncontrolled by the higher centers. An example of such difference in the response is that produced by pinching or pricking the skin on the back of the foot, or by pricking the sole of the foot. If the stimulus is relatively light it excites no response in the normal man, but when the cord is transected as mentioned above, there is a sudden withdrawal of the foot with dorsal flexion of the big toe, and with dorsal flexion of the knee and ankle, and with dorsal flexion of the knee and thigh. This is the defense reflex of withdrawing the member from an irritant, and it is typical of all cord injuries for the parts below the level of the injury. A similar response is found to a much less extent, however, in disease lesions of the corticospinal tract. Along with the flexion responses mentioned above, there is quite often an extension of the leg and foot on the opposite side, that is, with the foot and leg in extreme extension and the toes pointed downward. Such a response takes place in a more diffuse cord lesion and is known to physiologists as the *crossed extension reflex.* A modified form of the defense reflex of flexion is the well known diagnostic sign named after *Babinski,* which is a dorsal flexion of the great toe upon stimulation of the plantar surface of the foot and is diagnostic of some disease interfering with the corticospinal tract. Infants normally show this response until they are about one year of age, at which time it becomes plantar flexion rather than dorsal flexion and so remains throughout life, barring a corticospinal lesion.

In addition to the defense responses resulting from skin stimulation, the distal portion of the transected cord is capable of certain other deep reflexes, such as the so-called ankle-tendon and knee-jerk reflexes. In the case of the knee-jerk, the reflex is caused by sudden extension of the quadriceps muscle by tapping its tendon just below the patella. This sudden stretch of the muscle results in a myoclonic contraction of the quadriceps with an extension of the leg on the thigh. Stretching of almost any muscle by tapping its tendon will cause some jerk response. These responses have been called myotactic reflexes. Since the postural tension of the muscle is a function of the hindbrain it is usually considered that these centers in the hindbrain are over-active if there is an increase over what is considered to be the normal jerk response, probably due to a cutting off of control by the higher centers. This increased activity of the muscles is known as spasticity or increased muscle tonus. Strictly speaking, when using the word tonus in connection with muscle it should be qualified in terms of whether or not the muscle is static, or in a kinetic state, or whether the movement is slow or postural.

Special sense organs are also capable of eliciting "segmental" responses. For instance light thrown into the eye is transformed into nervous impulses and passes back as a nerve impulse over the second cranial nerve, and is reflexed back over the third nerve resulting in contraction of the pupil. This is known as the light reflex of the pupil. It should be noted in connection with the pupillary response to light that this is in part an autonomic response and is therefore rather sluggish.

In this connection it should also be noted that there is a rather definite segmental character to the arrangement of the nerves in the brain itself. For instance, consider herbivorous animals which seek safety in flight, such as a rabbit or a deer. Here the ears are large and movable, and the eyes are laterally placed, are rather large and command a very wide angle of view. *Dukes*[1] states that the divergence of the optic axes in the rabbit is eighty-five degrees. Such a divergence would permit a rabbit to see a virtual circle simultaneously, i.e., he could see straight back as well as straight forward at the same time. The visual axis divergence angle of the deer is just a few degrees less.

A grazing deer will graze with his nose up-wind, the olfactory bulb in the nose is the receptor for the *first* cranial nerve, by which means he would smell a predatory animal coming from the direction of the wind. At the same time his eyes, being laterally placed, could watch both sides and the rear. Here we have the receptor for the *second* cranial nerve. Any sound coming from the sides or rear would stimulate the eighth cranial nerve and would result in a reflex over the facial nerve with rotation of the ears so that they would face the direction from which the sound

[1] Dukes, "The Physiology of Domestic Animals," p. 593, Comstock Publ. Co., 1937.

came. Simultaneously, the sixth nerve reflex would tend to turn
the eyes in the same direction, thus giving the animal the advantage of seeing as well as hearing an approaching enemy.
These combined responses are known as the "sentinel response."
It is not a pure reflex, but requires certain short association paths
in the brain, yet it is essentially segmental in character, in that it
involves the sixth, the seventh and the eighth nerves.

In the case of a carnivorous animal the order of segmentation
is just a little easier to grasp because it does not involve a senti-
nal response. Let us follow a carnivorous animal in search of
food. First, by means of his olfactory sense over the first nerve
he scents or gets the scent of some animal which he can use for
food. Still using the first cranial nerve, he follows the trail until
eventually the food comes into sight, thus stimulating the second
cranial nerve. Carnivorous animals have their eyes set in the
front of the head and possess some stereopsis, and a fairly highly
developed convergence power. Sighting the animal immediately
activates the motor oculi brain segment activated by the third and
fourth nerves, so that he can judge the proper distance and position in space of the prey. When he gets within striking distance,
he pounces upon his prey and with the fifth nerve, which controls
the masseter muscles in the jaw, and also innervates the antenna
hairs on the face, whereby quick sensation is conveyed to the
brain centers, with a resulting smashing bite, usually at some
vital point. The animal is torn to bits by using the fifth nerve,
tasted by the ninth, and swallowed by the ninth, following which
the food passes into the stomach and comes under the influence
of the tenth, and perhaps eleventh and maybe the twelfth cranial
nerves. We thus see that the acts of securing and eating food, in
the carnivorous animal, becomes one of a seriatum response by a
certain segmental arrangement in the brain from before back-
wards. The following diagram illustrates the segmental arrangement of the brain somewhat schematically, page 49.

It is interesting to note, also, in this connection that pathological processes in the brain and brain stem seem to localize in the
brain segments as outlined.

There is, of course, the superior segmental mechanism which
operates by exercising some control, or some modification of the
more or less primitive reflex responses described above. This
superior segmental control will be discussed in a later chapter.

Whereas we have been discussing the brain as though it was composed of relatively independent functional centers, the fact remains that the brain functions as a unit and as a whole. The present trend of thought is away from the former belief that there were definitely localized centers within the brain. While this is true to some extent each of these centers and its activities are definitely inter-related. The *Journal of the American Medical Association*[2] has the following to say:

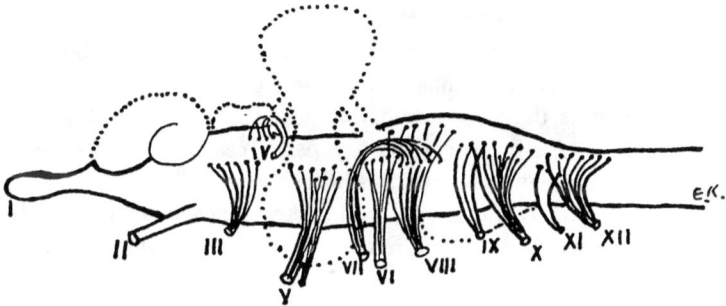

FIGURE I

The more recent studies, as Howell has pointed out, have tended to modify these extreme views as to localization and to emphasize the fact that histologically and physiologically the entire cerebrum is connected so intimately, part to part, that although the different regions mediate different functions, nevertheless an injury or defect in one part may influence to some extent the functional value of all other regions in the organ. The general idea of a localization of function has been accepted, but the modern view is that the cerebrum is composed of a plurality of organs, not completely separated one from another, as taught by Goll, but intimately associated and to a certain extent dependent one on another for their full functional importance. Dandy has extirpated both frontal lobes in a case of tumor, leaving the patient perfectly oriented as to time, place and person; the memory is unimpaired; he reads, writes and conducts mathematical tests accurately; his conversation is seemingly perfectly normal. Furthermore, by the excision in other cases of the left occipital lobe and of the lower third of the left temporal lobe,

[2] Editorial Journal American Medical Association, p. 95, 1268, 1930.

we can be sure that none of these regions are responsible for intelligence. . . . The entire body of the corpus callosum may be split in the middle without any appreciable disturbance of function. This structure is, however, eliminated from participation in the important functions which hitherto have been ascribed to it.

The intracortical connections between the several parts of the cortex, and the communicating paths to the diencephalon, function to a large extent as a result of the receipt of impulses coming over sensory pathways from the various end-organs of the body, such as in the skin, mucous membrane, special senses, and touch which latter is divisible into pain, heat, cold and light perception. These extero-ceptive sensory paths reach consciousness on two levels, first, in the thalamus where a certain crude integration, below purely conscious level, takes place resulting in diffused responses, and, second, in the cerebral cortex where a purely conscious perception takes place. These are the only two levels which have been determined experimentally, the former is necessary for the survival of the animal, while the latter makes him a conscious gregarious animal.

In studying sense perception great care is necessary to avoid the use of a stimulus which is much greater than is absolutely necessary to elicit response. Many examiners get faulty results and faulty data due to the fact that the stimulus they use is *too* powerful or *too* crude and may result in temporary shock to the receptor or its sensory path. For instance, passing a lead pencil lead over a skin area in an effort to determine tactile sense may cause enough stimulation, due to deep pressure, to result in a crossed response. This is not a desideratum. A true test of the ability of sensory tactile receptors and their fibers to function would properly be made by the use of a hair or a very small brush such as used for water color painting, and even then most of the hairs should be cut away. Such a device will measure tactile sense in the absence of pressure great enough to elicit deep pressure responses. It must be kept in mind constantly that for the detection, examination or measurement of the ability of a sensory receptor to function requires a *just liminal* stimulus and no more. The purpose of mentioning these facts is to impress the extreme delicacy of the responses, and to direct attention to the use, for the purposes of examination or therapy, of a minimal liminal stimulus, thus securing a maximal response, within the physiological limitations of the reflex arcs and the supersegmental influences brought to bear within the animal.

In an effort to show the wide ramification and the delicacy of stimulus required let us trace a few tracts in the brain. From the two maculae of the eye, in addition to the sensory optic nerve fibers, we have two sets of other fibers distributed to the vestibulo-cerebellar nucleus, to the lateral nucleus and to the vestibulo-spinal nucleus. This whole group of nuclei and tracts constitutes a single reflex unit. For practical purposes *Cobb*[3] says that there can be said to be five sets of entering fibers, five nuclei, and five tracts with the necessary neuro-mechanism to make them effective. The cerebellum, in which ends the vestibulo-cerebellar tract, may be functionally and anatomically an over developed vestibular nucleus. From the superior nucleus there is also a tract through the midbrain and thalamus, which latter may be the path which permits impulses to reach the conscious level, causing such sensations as dizziness, and perhaps nausea. The functional significance of these connections are such that stimulation of the vestibular end-organs in the ears causes impulses to pass to the five vestibular nuclei and spread into the medial longitudinal fasciculus and to enter the third, fourth, and sixth nuclei resulting in nystagmus. The impulse may even spread into the nucleus of the tenth nerve, which is distributed to the stomach and other viscera, and if there be such a spread there usually results the sensation of nausea. Actually, if the stimulus is great enough, there will be a spread into the nucleus of the eleventh nerve, which is distributed to the trapezius and sternomastoid muscles, and may result in some abnormal head posture. If the stimulation is still greater it may spread into the spinal connections resulting in a staggering gait, and clumsy hand movements. If the spread is into the cerebellum, from the vestibulo-cerebellar nucleus, then we will have exaggerated postural positions in an effort to maintain equilibrium. This whole picture is one of sea-sickness or car sickness, and may also be caused by rhythmic movements of the eyes, such as takes place during riding, or looking at moving objects, as well as by stimulation of the vestibular nerve in the ear by stimuli such as the rolling motion of a boat or movement

[3] Cobb, Preface to "Nervous Diseases," Wm. Wood, 1937.

by a train.   It should be noted in this connection that in the case
of true nausea, due to some irritant in the stomach, this reflex
works in reverse causing nystagmus, postural position changes,
and quite often causes a staggering gait, and crude hand movements.   The conclusion is that the thalamus and central brain
react in the same way, regardless of whether the stimulus comes
in over one set of sensory fibers or another.   It seems, therefore,
to be merely necessary that sensory impulses reach these centers
to elicit responses in their connected organs.

These examples of rather complex reflexes give an idea of the
general neurological structure which underlies the functions discussed.   It should not be thought, because these have been
studied separately, that there is any such definite separation in
the nervous mechanism of the animal, because such is not the
case.   The whole nervous system acts as a whole and all of its
parts are more or less dependent upon each other.   The only purpose in studying these divisions is that they give a better insight
into the functions as a whole and aid in forming a fairly clear
picture of the function of some of the parts.

In the foregoing we have presented some of the phases of reflexes as they take place within the neurone and the association
paths.   There is another type of response of the nervous system
which has many of the elements of the reflexes, but is in no sense
a true reflex.   This is the conditioned response.   A conditioned
response may be developed by consciously performing an act in
association with a given experimental pattern.   *Pavlov* was the
first to reduce the study of the conditioned responses to a practical means of investigation.   One of his famous experiments consisted of sounding a bell simultaneously with the delivery of food
to the dogs in the cages.   After this procedure had been repeated
daily for a number of days, a point was reached where it was only
necessary to sound a bell to produce excessive salivation by the
dogs used in the experiment.   As a result of a continued sounding of the bell, in the presence of food, some association tract or
path had been developed within the nervous system of the animal
such as to cause it to respond physiologically to a stimulus which
ordinarily would *not* have called forth such a response.   Any
sensation can be used to condition a response, such as salivation
in the experiment mentioned above.   It is only necessary that the
new sensation be repeatedly associated with the response desired.

These responses are sometimes called habit responses, and as the Duke of Wellington facetiously remarked, "Doing a thing twelve times forms a habit." This may or may not be strictly true, but it is a fact that acts which can consciously be performed by an individual will usually establish a conditioned response that reacts like a reflex, following the meticulous execution of a given act for about twelve or fifteen times. Of course, in this latter instance, we are dealing with striped muscles, because these are the only ones under conscious control. Glands and plain muscle are under the control of the autonomic, and hence beyond the power of the cerebrum or pure cerebration to set up conditioned responses by conscious effort. If conditioned responses are to be set up in glands or in smooth muscle, the stimulus or sensation must be a simultaneous one and be synchronized with some normal physiological process inherent in the autonomic.

A conditioned response remains permanent *only* if the incentive for its performance continues, or if it is repeated at rather regular intervals. In other words, a conditioned response tends to revert to its ordinary unconditioned status, unless it is constantly being reconditioned. *Culler.*[4]

In the central gray in the vicinity of the nuclei mentioned above, particularly the vestibulo-cerebellar nucleus and its mediated control of muscle tonus, both postural and to some extent static, is found the respiratory center which controls the rate and depth of the act of respiration. *Henderson*[5] has shown that this muscle tonus, controlled by the central nervous system and influenced by the respiratory center in the brain, is the basic factor which determines both the amount of metabolism and the corellated volume of circulation of the blood and the volumetric respiratory capacity of the individual. In other words the dominant factor in circulation of the blood becomes one of the tonus of the muscle involved in the act of respiration, which is itself controlled by a mass of nerve cells in the central gray. Incidentally, this mass of nerve cells is chemically controlled, being activated by the proportion of carbon dioxide in the blood stream. Here is seen an interlocking of blood chemistry and the nervous system. Many other such instances are also known to exist, some of which will be mentioned later.

[4] Culler, Univ. Ill., reported S.N.L., Dec. 21, 1935.

[5] Henderson, "Mechanical Factor in Respiration," Science, 79, 2057.

CHAPTER VI

CENTRAL GRAY

In the preceding chapter mention was made of that portion of the brain which is often spoken of as the "central gray." Another common name is the archipallium, indicating that it is an old brain, i.e., phylogenetically older than the other portions of the brain. It is doubtful if it can be successfully shown that, in the process of evolution, the central gray antedates the cerebrum or cortical structure. Both are found in all of the lower orders of animals as far down the scale, and as far back chronologically, as it is possible to secure accurate information. It may be that the term implies the control of the more primitive phase of living and if this be the usual implication of the term, undoubtedly the central gray reigns supreme in its field. By primitive functions is meant those functions having to do with the control of, maintenance of, and expression of life.

Properly to understand the central gray, here used in its broadest sense, it seems imperative that the embryology of its formation be first discussed. In the early development following fertilization of the ovum, there is found on its surface the embryonic plate or area. This area is the first mass of cells to undergo division under the stimulus of impregnation. The area takes the rough outline of a pear-shaped body, its thickness is probably one-sixth of its length. Down the center of this mass extends a depression called the neural-groove. This is the first differentiation of tissues in the embryo. *Morris.*[1] This embryonic mass consists of three rather loosely divided layers of cells in the tissues forming it, having the names *ectoderm* for the external layer, *mesoderm* for the middle layer, and *entoderm* for the inner layer. The skin and nervous system develop from the tissues known as the ectoderm. The phrase "nervous system" is understood to be and to include the brain, nerve cells, axones, dendrites, synapses, and all other tissues having that degree of irritability peculiar to this system. Incidently, since ocular problems are to be discussed later it might be well here to mention that the lens, iris and cornea of the eye develop from the epidermis which is originally a part

[1] Morris, "Morris' Anatomy," p. 753.

of the ectoderm, in the early developmental stage, as cited by *Morris*.

The neural groove, by a process of invagination, deepens into the embryonic area and then closes until it becomes a tube, known as the neural tube.

As soon as the neural tube in the embryo has taken a definite form, expansions develop at its anterior end, alternated with two constrictions, forming three vesicles, known as the anterior, middle, and posterior vesicles. At this stage we have mere tissue bounding the neural tube, which envelops a liquid, hence the term vesicle. As the embryonic area and neural tube continue their development a diverticulum moves posteriorly from the median line of the anterior vesicle until it almost contacts the constriction at the anterior border of the middle vesicle. As soon as this process has been completed the walls of the vesicles undergo a thickening process, which thickening when completed constitutes the brain and its several parts.

The thickening walls around the original posterior vesicles become the rhombencephalon which, later in life, has largely to do with certain coordinational activities. The walls of the originally middle vesicle develop into the mesencephalon and the corpora quadrigemina, otherwise known as the colliculi.

The walls of the anterior vesicle, following its division into two vesicles by the diverticulum mentioned above, develop three ways, or rather, into three parts: The two cerebral hemispheres; and the diencephalon, the latter being sometimes called the interbrain, but, in connection with certain other closely associated parts, we shall speak of it as the "central gray." The cerebral hemispheres are also known as the telencephalon, presumptively because it is this portion of the brain which enables the animal to perceive and evaluate things not within his own skin, i.e., at a distance. The walls laterally to the mesial diverticulum send other diverticula forward on each side which, as development proceeds, eventually coming to an end in the nasal cavity. In man this ending is in the Schneiderian membrane, having to do with the sense of smell. When this development is completed the diverticulum becomes known as the first cranial nerve.

Chronologically speaking, just prior to the beginning of the development of this olfactory mechanism two vesicles form on the lateral walls of the originally anterior vesicles. These vesicles are the optic vesicles, and by advancing laterally, followed by invagination of their distal ends, result in the formation of those parts of the eyeball posterior to the limbus, sclera, choroid and particularly the retina. Incidentally, the optic vesicles and visual mechanism is the *first* semblance of development of specialized tissue in the body and becomes the sense organ which is supreme in the maintenance of the safety of the animal. Chronologically, the olfactory mechanism begins development just subsequently to that of the optic mechanism.

These portions of the central gray which develop from the posterior portion of the originally anterior vesicles are: The two masses of the thalamus which lie laterally to the third ventricle, originally the middle vesicle, the geniculates, lateral and the medial on each side; the hypophysis or pituitary which projects downward, and anteriorly from the floor of the anterior portion of the third ventricle; the epiphysis or pineal gland posteriorly; and the corpus striatum laterally to all of these. The corpus striatum is important in life later because it is that portion of the brain which controls the muscle tonicity of skeletal muscles throughout the body, and any failure or involvement of this region results in nystagmus, agitans and other incoordinate and involuntary motions. In connection with the corpus striatum it has been pointed out by *Ranson,*[2] and others, that this mass of nerve cells is directly controlled by the thalamus. *Tilney and Riley*[3] and others confirm the findings in this connection cited by *Ranson, Britton, Becterev, Dana, Cannon,* and others all locate the control of the vital and defensive mechanisms of the body in the diencephalon and the mesencephalon. *Tilney and Riley* also include the nucleus ruber in this connection. The mechanisms whereby the vital and defensive control of the body takes place will be rather fully considered elsewhere.

It should here be recalled that the corpora quadrigemina, otherwise known as the colliculi, developed out of the originally middle vesicle.

Cannon[4] states that the diencephalon is the coordinating center

[2] Ranson, ''Anatomy of the Nervous System,'' p. 319, Saunders, 1927.

[3] Tilney and Riley, ''Form and Function of the Nervous System,'' pp. 607–825, Hoeber.

[4] Cannon, ''Wisdom of Body,'' pp. 198 and 208.

for the sympathetic and the autonomic and indicates that a breakdown in this coordinating function will actually endanger life by the destruction of these processes which maintain it. It must here be pointed out that all of this control is quite aside from the functioning of the conscious mind of the individual. The control of these associated functions for the maintenance of life cannot be attributed to the diencephalon as a prime act, because the true *modus operendi* is *reflex* and depends upon the receipt of afferent impulses by this region, for the moment, leaving out of consideration the source or kind of stimulus which initiates these impulses. This latter statement imples that these centers in the brain function as do ganglia, in that upon the receipt of a proper afferent impulse, the efferent impulses are released to the organs and tissues which severally respond in terms of the then needs of the body, whether they be chemical, physical, mechanical, or otherwise.

Returning to the eye, and its connecting nerves, we find that the sensory paths from the retina, behind the decussation, are delivered to three places:

1. The thalamus, as cited by *Haggard*[5] in reference to the function of sight.

2. Another well recognized distribution is to the anterior colliculus, otherwise known as the superior corporus quadrigeminum. This center is the reflex center of the eye, largely mediated through the tectospinal tract, *Rasmussen*.[6] *Wernicke* variously quoted by the authorities, holds that the superior corporus quadrigeminum is not involved in the act of seeing as such, but that through the tectospinal tract it involves parts below, due to the white fibers entering the central gray of the cord at the several levels. *Tilney and Riley*[7] state that both the pulvinar of the thalamus and the superior corpora quadrigemina send fibers to the spinal accessories which has connecting links with the tenth cranial nerve.

3. *Herrick*[8] has shown that there are definite cell groups in the lateral geniculate bodies which are connected directly with rela-

[5] Haggard, ''Anatomy of Personality,'' p. 81, Harper and Bros., 1936.

[6] Rasmussen, ''Principle Nerve Pathways,'' pp. 28–29, MacMillan, 1932.

[7] Tilney and Riley, supra, p. 782.

[8] Herrick, University of Chicago, Science, No. 10, 1933.

tively fixed locations of the retina. It has long been recognized that the geniculates, probably mediated by way of Gudden's commissure, are reflex centers, or the centers capable of conditioning which enable the eyes to fixate the same object simultaneously. *Villiger*[9] mentions nerve fibers leaving primary centers and passing anteriorly forward to the retina. The function of these fibers is obscure, although they have been called "retinomotor fibers"[10] probably not because they produce motion in the retina, but because they are efferent fibers. Later the writer will suggest a possible function of these fibers in connection with the development of his thesis under the heading "Ocular Control of Supportive Functions."

The superior corpora quadrigemina also receive fibers from the occipital lobes in the brain, *Morris.*[11] Some authorities, due to this fact, attribute to them the production of the systemic responses which result from visual impressions which have reached the visual centers in the occipital lobe, either as pure reflexes or in connection with some emotional response thereto on the part of the subject.

Histologically, the central gray is made up of cell masses and nuclei composed of nerve cells whose dendrites lock and interlock throughout the masses and through which connecting paths may be traced to the various parts. It should also be mentioned that there may be and probably are purely synaptic paths through this region, due to having been repeatedly traversed by impulses set up either in the periphery, viscera or other sensory receptors. Synaptic paths thus used, over considerable periods of time, have their liminal value so lessened that a wave of impulses through the central gray tends first to excite those cell groups having the lowest liminal value.

With this brief picture of the central gray kept in mind it should be easy to follow a more or less detailed description of functional control localizing in this region. *Pflüger*[12] laid down the principle in 1877 that "the cause of every need of a living being is also the cause of a satisfaction of that need." *Fredrics*[13]

9 Villiger, "Brain and Spinal Cord," p. 172, Lippincott, 1918.
10 Starling, "Human Physiology," p. 42, Lea and Febiger, 1933.
11 Morris, supra, pp. 821 and 866.
12 Pfluger, "Pfluger's Archives," 1877, XV, 57.
13 Fredrics, "Archive de Zoological Expe. et Gen.," 1885, III, 35.

states that every living agency sets up a compensatory activity in an effort to neutralize or to repair disturbances or damage resulting from environmental contact or changes. These compensatory changes have for their purpose the maintenance of a state of physico-chemical equilibrium within the organism. This equilibrium might well be called a syntonic state of the organism and is maintained by the coordinated functions of the several parts, to wit, i.e., brain and nerves, heart, lungs, liver, spleen, endocrines, all working cooperatively. The reactions that are inherent, also known as the instinctive type, have their central control in the central gray in the basal part of the brain. In fact, they will maintain syntony by compensation and by altered function, even if all parts of the brain anteriorly and superiorly to the diencephalon are ablated, as demonstrated by *Bard*,[14] and confirmed by others. However, this functional control, and the phenomena associated therewith, completely disappear following destruction of the diencephalon. This region of the brain receives impulses from the many receptors throughout the body, both proprioceptor and the spatial receptors such as the senses of sight, smell, hearing, taste, and touch, and upon their receipt it functions reflexly *without* the aid of the cortex.

Among the vital function centers known to be localized in the central gray are:

1. The respiratory center, which is a nucleus that controls the rate and depth of respiration, but which itself is chemically controlled by the carbon dioxide content of the blood;

2. Another center known to be localized in this region is that which controls the constancy of the body temperature, i.e., a thermostatic effect. The end-result probably is an admixture of two mechanisms: One is a control of the rate of metabolism, including the oxidation of blood sugar by the tissues, and the other, a control of the heat radiation through the skin by either a vasoconstriction or vasodilation depending upon the need of conserving heat or ridding the body of an excess;

3. Another center in this region is the glycogenetic or glycometabolic center. Overactivity of this center, through its effect upon the adreno-sympathetic stream, which excess, if not adequately taken care of by the pancreatic secretion of insulin, re-

14 Bard: American Journal Physiology, 1928, LXXXIV, 490.

sults in the disease known as diabetes mellitus. *Kirby*[15] has shown that an emotional shock may cause this excessive release of blood sugar resulting in a hyperglycemia which may not necessarily be the result of a failure of the pancreas properly to aid in the metabolism of sugar. In other words, continued emotional stress and shock may cause diabetes in an individual whose nervous system is so constituted that the sympathetic response does not return to its normal in a reasonable length of time following an emotional shock. A similar situation exists in practically any of the other functions having localized functional control in the central gray.

4. Yet another center in this region is the cardiac center which, mediated through the vagus nerve, inhibits that organ slowing its rate of beat, but increasing the force simultaneously. There exists a well known oculo-cardiac reflex which may be used to demonstrate this response, and also proves a connection between the eye and the functions controlled by this region, mediated through the autonomic nervous system, which will be discussed later.

5. Still another center in this region is the vasomotor center, which, either directly or mediated through the adrenals, may constrict the arteries thereby raising blood pressure. Another function attributed to this particular center is that of producing a compensatory vasodilation in some part of the body remote from any region which may be in excessive vasoconstriction, thus maintaining a fairly average level of intra-vessel blood pressure.

Vanderahe[16] localizes certain centers in the central gray in reference to the eyes themselves. To those for the diencephalon proper, consisting of the sub-thalamus, the thalamus, the epithalamus, and the hypothalamus, he attributes faults of the extrinsic eye muscles, optic radiations and the optic chiasma, while in the midbrain he localizes faults of lid muscles, and the pupil. But since the pupil and the ciliary muscle receive their supply from the ciliary ganglion it seems reasonable that he should have included accommodative faults in his list. But *Barenne*[17] re-

[15] Kirby, "Emotional Disturbances Cause Physical Diseases," p. 364, S.N.L., January 9, 1934.

[16] Vanderahe, "Pathological States of the Nervous System and Eye," Ohio State Med. Journal, Feb. 1933, p. 110.

[17] Barenne, paper before A.A.A.S., 1937.

ported, in a paper read at the 1937 meeting of the American Association for the Advancement of Science, that the sensory-motor cortex and the visual cortex were definitely related to the thalamus as determined by electrical recording instruments in conjunction with localized use of strychnine. *Lashley* at the same meeting in a paper, ''Inter-Play of the Cortex and Thalamus in Sensory Functions,'' reported new evidence which showed that the subcortical visual mechanism is much more important in mediating animal and biological behavior than had been known prior to February of 1937. Also *Stephens* simultaneously reported definite evidence that the cardiac response from this region is due to activity of the para-sympathetic system which has its major origin in the central gray. *Finkelman and Haffron*,[18] confirmed by *Ransom*, cite evidence to prove that the hypothalamus in the central gray is the primary factor in the mental disease known as schizophrenia. Also that the hypothalamus controls oxygen consumption of the body and other mechanisms which are known to be disturbed in that disease. *Mason*[19] states that the neuro-physiological processes of the cerebral cortex are geared to those of the thalamus, and through it to the entire nervous system, to the endocrine glands and to the whole animal. *Papez*[20] states that consciousness of emotional response seems to be mediated as an end-product by the diencephalon, stating that the sense impressions reach the hypothalamus, emerge through the mammillary body and pass to the cortex by way of gyrus cingulus.

Among conditions that are known to be produced by faulty function of the hypothalamus are diabetes insipidus, polyurea, polydypsia, cachexia, genital atrophy, adipose genital dystrophy, convulsions, glycosuria, altered metabolism, faulty visceral function without structural change, faulty and altered fat, protein, carbohydrate, mineral and water metabolism, temperature regulation, and endocrine regulation. *Karplus and Kareidl*,[21] confirmed by *Beatti* demonstrated that stimulation of the posterior

---

[18] Finkelman and Haffron, Science, 85, 2193, 10.

[19] Mason, ''Science and the Rational Animal,'' Science, 84, 2169, 73.

[20] Papez, ''Proposed Medium of Emotion,'' Arch. Neuro. and Psych., 1938.

[21] Karplus and Kreidl, quoted by Vonderahe. ''Representation of Visceral Function in Brain,'' Ohio St. Med. Jr., Feb. 1935, p. 107.

hypothalamus would raise blood pressure in an adrenal-ectomized animal, and in addition thereto would cause dilation of the pupils.

All cranial nerves arise from or enter the central gray. The two most highly organized paths carrying the impulses into this region are the optic and olfactory paths. There is a direct connection between a third afferent path entering this region, the auditory, and the optic tract by way of Westphal's ganglion. *Hartman*[22] reports as the results of experiments, conducted by the Pennsylvania State College, in which it was found that simultaneous stimulation of the sense of smell was an aid to seeing. His experimental reports do not indicate that if both the sense of smell and hearing were simultaneously stimulated there would be a further increase in acuity beyond that obtained by the stimulation of either of these alone. *Nugent*[23] ran a series of experiments in which about eighteen odors which the subjects inhaled, were used in an effort to determine their effect upon the sense of sight. He found that the odor of oil of citronella gave a maximum increase in acuity discrimination, while the odor of turpentine almost invariably decreased acuity in the subjects used in his experiments.

A ganglion may roughly be defined to be a mass of nerve cells which upon receipt of afferent impulses relays a response over some one or more efferent paths leading from the cell mass. However, it is not known if a ganglion possesses discriminatory power in reference to any specific distribution of the efferent train of impulses. It is known, though, that in large masses of cells, such as the central gray, there are numerous nuclei to which can be attributed specific functions. These nuclei usually have a considerably lowered threshhold value than that of the general mass of cells. When the waves of energy, which constantly traverse the mass, are in any way altered by the receipt of impulses by the mass, there does come into action a relative selectivity which determines where the major efferent response will be manifested, due to the lowered liminal value of some of the constituent nuclei within the masses. The thalmus and central gray are just such a mass of nerve cells.

The thalamus and central gray receive afferent tracts from the olfactory, optic, auditory and all ascending tracts from below.

22 Hartman: ''Ears and Nose and Sight,'' S.N.L., Feb. 11, 1933, p. 86.
23 Nugent, Syntonogram, C.S.O., April 1935.

These tracts from below are largely proprioceptive afferent paths. *Kirke*,[24] *Morris*,[25] *Dana*,[26] *Cannon*,[27] and others are in agreement that there are no afferent paths passing into the brain that do not have direct or relayed connections in these masses.

The thalamus and central gray receive or transmit all motor tracts leaving the brain for parts below, except the crossed pyramidal tracts, which are purely voluntary tracts from the cortex and activate volitional movements of the soma. *Kirke*[28] and others heretofore mentioned, have demonstrated these motor tracts to all of the viscera. Incidentally, in this connection, the thalamus, due to its cortical connections, can, under some conditions, seize control of the pyramidal paths and other normally volitional tracts of the brain. These connections account for certain automatic defensive responses by the animal, the changes in facial expression under emotional stresses and the apparently purposeful movements which usually require volition. These thalamic patterned responses, particularly those governing facial expression, are unique in that it is next to impossible voluntarily to assume facial expressions identical with those produced by thalamic control.

Tilney and Riley,[29] confirmed by *Howell*,[30] cite another most interesting fact in this connection, to the effect that while the corpus striatum does govern muscle tonus of somatic muscles, the striatal bodies themselves are under the *control of the thalamus*. Faulty action of the striatal body, thalamus controlled, may result in muscular atony, tremor, nystagmus, and clonic spasms of striated muscles, which latter classification also includes the six extra-ocular muscles in each orbit. *Howell* also indicates that the efferent paths from the corpus striatum bring it into connection with the viscera as well as with the musculature of the body.

As long ago as 1911 *Head* and *Holmes* showed that the thalamus was the center for pain perception of the body, regardless of the cortex. *Ranson*,[31] in this conection, states that it is important to

24 Kirke, ''Physiology.''
25 Morris, supra.
26 Dana, ''Diseases of the Nervous System.''
27 Cannon, ''Bodily Changes in Fear and Rage.''
28 Kirke, supra, p. 694.
29 Tilney and Riley, supra, p. 825.
30 Howell, ''Textbook of Physiology,'' pp. 239–240, Saunders, 1933.
31 Ranson, ''Anatomy of the Nervous System,'' p. 305, Saunders, 1927.

bear in mind that it is *not* necessary for pain impulses carrying painful impressions to reach the cortex before one becomes conscious of them, and that the *thalamus alone is sufficient for the perception of pain*. *Hess*[32] states that stimulation of the thalamus produces an inhibitory action by the thalamic centers on the cerebral cortex, resulting in sleep. The method of stimulation of the thalamus, used by *Hess,* is such a one that its continued application would result in definite inhibition of the nerve cells in its mass. Such an inhibition would depress the sympathetic and lessen its pattern-action effects, while at the same time it would relatively increase the parasympathetic control, and most authorities are now agreed that there is a very definite sympathetic control center in the central gray.

In addition to the foregoing, mention here should be made of the fact that the superior corpora quadrigemina receive fibers from the occipital lobe of the cerebrum. These fibers are the probable means of mediating pupillary contraction in the presence of excessive illumination, because it is known that the superior corpora quadrigemina control the well known parasympathetic act of pupillary contraction as cited by *Pottenger*.[33]

Summarizing the preceding we have here a picture of a "master ganglion" which receives all afferent impulses, including painful ones, and discharges all efferent responses, except those which are distinctly volitional, yet under conditions of great danger the thalamus and central gray may seize control of even these. *Cannon*[34] says that "the thalamic patterned processes are inherent in the nervous organization, they are like reflexes in being instantly ready to seize control of the motor responses, and when they do so they operate with great power." In this connection it might be well to mention that a few of these somatic motor responses can be controlled to some extent by conditioned reactions of cerebral origin, as a result of previous experiences or impressions by the individual. This applies to the peripheral machinery, the viscera excepted.

The following known expressions of function are all possible in the decerebrate animal: sneezing; licking with the tongue; eating, digestion and assimilation and elimination; respiration; vomit-

---

[32] Hess, Lancet, 1932, 223, 1259.

[33] Pottenger, "Symptoms of Viscera Diseases," p. 248, Mosby.

[34] Cannon, supra, p. 374.

ing; facial reflexes by tickling hair in the nose; sexual mating and reproduction; lactation; running; stepping; rebound extension; and withdrawal of the extremities from an irritant. Should the nucleus ruber be damaged, in the mutilating operation of removing the cerebrum, there will also be decerebrate rigidity in extension.

Electrical and mechanical stimulation has demonstrated the existence of an important group of autonomic centers in the lower part of the diencephalon at the base of the brain. The functions of these centers may be summarized as follows:

a. Thyroid gland stimulation
b. Lacrimal gland stimulation
c. Sugar metabolism regulation
d. Adrenal gland stimulation
e. Salivary gland stimulation
f. Kidney regulation
g. Sweat gland stimulation
h. Vasomotor control
i. Fat metabolism regulation
j. Possibly uterus control
k. Possibly bladder control
l. Temperature regulation
m. Sebaceous gland stimulation
n. Protein metabolism regulation
o. Pupil regulation

From the foregoing it can be seen that the animal, although unconscious, due to a lack of a cerebrum, still is capable of living; reproducing its kind; partaking of and utilizing nutrition; ridding himself of waste, ridding himself of irritants in the nose or bronchi by sneezing or coughing; ridding himself of irritants in the stomach by vomiting; or if the irritant be in the intestines, eliminate it by diarrhea. Also, he will protect his extremities from damage by making an effort to remove them from the immediate presence of any irritant. It should also be noted that this animal is still capable of walking, running, and the usual coordinate movements associated therewith, this coordination is, however, mediated by the cerebellum.

Edinger and *Fisher*, in ''Pfluger's Archives'' report the instance of a child born decerebrate who lived to the age of four years. In 1928 *Dr. Baker* of Ohio State University, department of Anatomy, showed the writer a brain shipped in for minute examination from a western state.   This brain was removed post mortem from an individual thirty-three years of age, and macroscopically showed no part to which could be assigned cerebral or cortical functions.   These are two instances of known decerebrate life and function in the human being not produced experimentally.   The conclusion reached is that the cortex is *not* necessary for the maintenance of life, reproduction, and a reasonable protection and defense of the organism.

It follows, therefore, that the cortex plays no part in the reflex control or activation of these functions.   On the other hand, *Cannon*[35] insists that the cortex cannot prevent thalamic patterned action, and that the viscera are only under thalamic control.

It is interesting to note that the thalamus always reacts in the same way, via the sympathetic, when called upon by any circumstance requiring defense response by the individual.   There is *no variation*.   The end response is predictable with uncanny accuracy due to the fact that this "master ganglion" has a job to perform which phylogenetically is much older than man, and it has learned its lesson well.

Normal body functions, and normal ocular functions, depend upon the supportive functions thereof, the maintenance of which have been shown to be the particular duty of the thalamus and central gray.   A proper balance between the driving force of the function, and the restrainer of that function, is necessary for their proper manifestation.   How this balance, i.e., syntony, is mediated will be discussed later.   The normal body functions have rightly been called the "vital functions," and may be listed as follows: mental, motor, reflex, sensory, trophic, vasomotor, secretory, hormonal, and body temperature control.

A logical conclusion to be drawn is that if a means be discovered or developed for controlling the mass responses, and the pattern responses, of the thalamus and central gray, it should be possible to assure proper function of all of the above mentioned vital processes.   The eyes are known to have a direct connection with the central gray and may be used to exert a powerful influence thereon as will be discussed in a later chapter, at which place the proper authorities will be cited.

[35] Cannon, supra, p. 251.

CHAPTER VII

THE AUTONOMIC

In this chapter we shall investigate the equilibrium, or Syntonic, tendency of the vegetative nervous system. It is this system which maintains life in the individual, the processes of which may be mediated through the functional control of body chemistry mediated by the endocrine glands, or by direct control of the organs and larger tissue masses of the body as distinguished from cell function *per se*.

This act of living is roughly described by the word metabolism, a process embodying two antagonistic functions, namely, anabolism or the up-building process, and catabolism, the destructive or down-tearing process. When these two processes are properly integrated and balanced we have the physiologically normal individual. Anabolism is purely that phase of metabolism which might be called the vital phase, meaning that phase which is necessary for the ultimate maintenance of life. Obviously, this phase embraces the digestion and assimilation of food and its distribution by the blood through proper circulation. It should here be understood, however, that the cell itself must make proper use of the food and oxygen supplied to it by the circulatory mechanism if cellular, tissue, organic or systemic function is to be maintained.

Catabolism, on the other hand, has to do with the breaking-down of foods by the vital processes within the cells themselves. This may take place during their normal life habits, or as a result of excessive strain or over-activity by the cells during functional activity in the presence of some emergency requiring a more marked function of the defensive mechanisms. Emergency functions are those which are peculiarly adapted to protecting the individual in the presence of danger or as a result of rage, resulting from circumstances which may jeopardize the continued life of the individual following contact with individuals or other forces in its environment.

This whole process of metabolism is under the control of, and dominated by, the autonomic nervous system, sometimes called "the vegetative nervous system." By autonomic is meant "self-

governing'' or ''self-sufficient,'' in that this system is functionally divided into two parts which are antagonistic each to the
other.   Not only is the vegetative system functionally divisible,
but in terms of the origin of the nervous trunks compromising the
system, it is anatomically divisible into two parts.   That division
having to do with the anabolic processes is variously named the
parasympathetic, or the craniosacral division.   This division has
its nerve trunk origins in the central gray surrounding the region
of the third ventricle in the brain.   This region in part is known
as the diencephalon and extends from the optic chiasm anteriorly,
to the epiphysis posteriorly, and from the thalamus superiorly

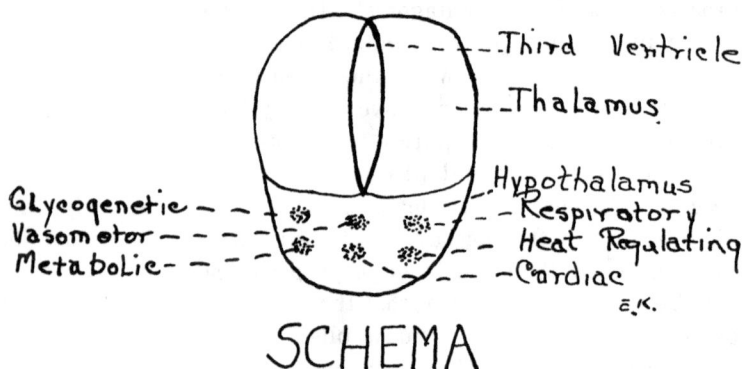

SCHEMA

FIGURE II

and through to the copora mammallaria inferiorly.   This region
has many detailed nerve fiber connections which have been set
forth rather minutely by *Jelliffe*[1] and *Kuntz*,[2] and will be briefly
detailed hereinafter.   This region, particularly the nuclear groups
in the hypothalamus, is in intimate connection with the pituitary,
otherwise known as the hypophysis, by nerve fiber connections in
addition to a specialized ''portal circulation,'' *Vonderahe*,[3] interlocking the blood supply of this portion of the brain with the
pituitary.   Figure II is a schematic representation of the centers
in this interbrain.   It must here be pointed out that this interbrain is made up of nerve cells whose dendrites interlock forming

[1] Jelliffe, ''Diencephalic Vegetative Mechanism,'' Arch. Neurol. and Psychiat., 32, 982, Oct. 1934.

[2] Kuntz, ''The Autonomic Nervous System,'' Lea and Febiger, 1934.

[3] Vonderahe, ''The Representation of Visual Function in the Brain,'' Ohio
State Med. Jour., 31, 104, Feb. 1935.

a veritable feltwork in the mass. This interlocking is such that an afferent impulse reaching the diencephalon will be transmitted throughout the entire mass and all of its contained nuclei. The result of this arrangement is that those nuclei and functional control centers within the diencephalon which have a lessened liminal or threshhold value will be the ones which will first respond over efferent paths, following the receipt of an afferent impulse. Out of this region and the midbrain we have the parasympathetic outflow by way of the third, seventh, ninth and tenth, and one or two authorities include the eleventh, cranial nerves. This parasympathetic outflow, in conjunction with the sympathetic proper, innervates all the viscera in the body. These two systems are so set up that they antagonize and check each other somewhat after the manner of the engine and brakes in an automobile. The parasympathetic activates some visceral functions, while slowing others. The sympathetic, always antagonizes the action of the parasympathetic, for any given viscus. Should there be any difference in the tension existing between these two divisions, there will be an alteration of function of the viscera in the direction of the more active division. In cases where the entire parasympathetic is dominant, stimulated or over-active, the individual is said to be vagotonic, whereas if the sympathetic should be dominant, stimulated to overactivity, the individual in this status is known as a sympathicotonic. For the purpose of this thesis, the diencephalon is by far the more vital, because it is the center through which emotional states are physiologized or effected. *Müller*[4] states that the central gray matter surrounding the third ventricle is the "seat of the elementary, vital and vegetative functions of life." On both sides of the third ventricle lies the masses of the optic thalamus as heretofore stated. Phylogenetically, the hypothalamus is one of the older and more constant findings in all of the orders of the vertebrates and is consequently very important in terms of specialized functions, having their nuclear regions therein. This region is sometimes called the archial portion, also known as the archipallium.

That division of the autonomic which antagonizes the parasympathetic, as has been said before, is known as the sympathetic and consists of two gangliated chains within the trunk which receives

[4] Müller, "Die Lebensnerven," Berlin, Julius Springer, 1924.

white rami communicantes from all the dorsal spinal segments, and usually the first three lumbar nerves. It sends gray rami to the cord at practically all levels, beginning with the superior cervical ganglion which sends gray rami to the first four cervical segments. Despite this fact the sympathetic nervous system has no segmental arrangement within itself and whatever segmental responses result from stimulation of this division are due to the segmental arrangement of the cerebro-spinal system.

The inter-communication of the sympathetic with the cerebrospinal system is widely distributed. The accomplishment of this distribution and integration requires that these impulses will traverse the spinal cord. However, no special sympathetic tracts have ever been identified in the cord as shown by *Thames*.[5]

Figure III schematically shows the autonomic connection to the viscera. In order to secure a clear picture of the functional control by the vegetative system, it seems wise to here list the respective controls under the proper divisions.

Hereditary dominance of, stimulation of, or overactivity of the parasympathetic, which incidentally includes the sacral nerves in addition to those heretofore named, does the following things:

Contracts pupil
Widens eye-slit
Increases lacrymation
Upper lid ptosed, puffy
Intra-ocular hypotension
Increases accommodation
Causes esophoria,—reflex
Low abduction tendency
Activates intrinsic eye muscles
Increases secretion of nose, mouth and pharyngeal glands, producing so-called catarrh
Increases salivary secretion
Hypersecretion and hypermotility of intestines leading to colicky pains and either spastic constipation or diarrhea
Slows heart
Dilates arteries
Decreases blood sugar

[5] Thames, "Allergy and Neuresthenia," Med. World, 57, 1, 38, Jan. 1939.

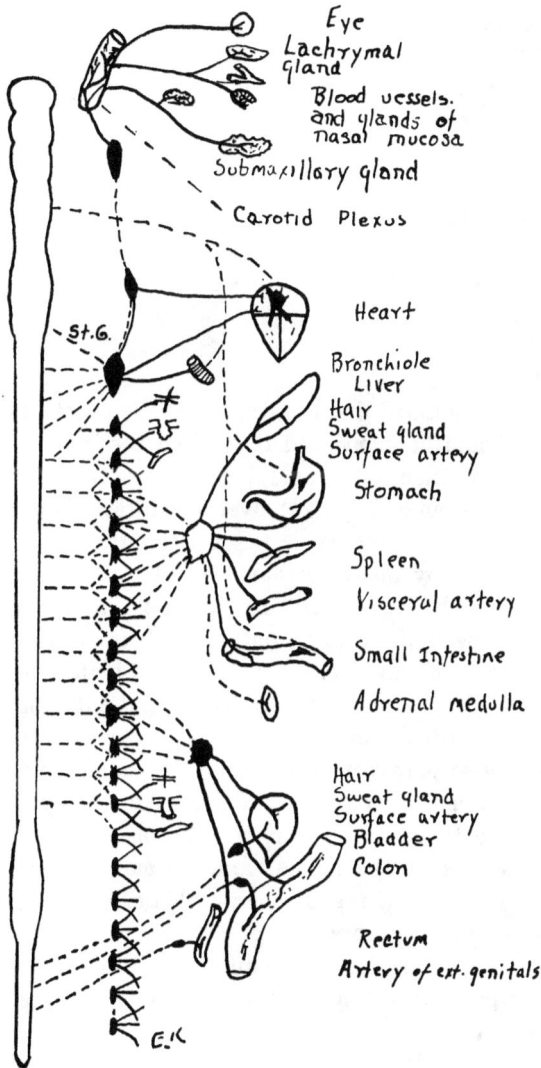

Eye
Lachrymal
gland
Blood vessels.
and glands of
nasal mucosa
Submaxillory gland
Carotid Plexus
Heart
Bronchiole
Liver
Hair
Sweat gland
Surface artery
Stomach
Spleen
Visceral artery
Small Intestine
Adrenal medulla
Hair
Sweat gland
Surface artery
Bladder
Colon
Rectum
Artery of ext. genitals

St. G.

E.K

FIGURE III

Stops sweating—palms, soles, under-arms
Decreases respiration rate
Causes irritable bladder
Decreases adrenalin—presumptively
Increases bronchial secretion

Produces asthma

Causes hypermotility and increased secretion of stomach
with an excess of hydrochloric acid.

Activates:

Parathyroids

Adrenal cortex

Stomach

Liver

Pancreas

Spleen

Duodenum

Intestines

Be again reminded that these are anabolic functions. Furthermore, it should be mentioned that when the individual is in the
horizontal position, such as lying in bed, this division of the
autonomic becomes dominant over its antagonistic sympathetic
division. This dominancy is taken advantage of by physicians
during illness, by insisting that the patient go to bed and stay
there until such time as the toxemic sympathetic symptoms have
been overcome by the parasympathetic. This handling of patients
takes the form of an axiom often mentioned by teachers of clinical
medicine, ''during an acute illness the patient should be transformed from a 'column' to a 'beam.' '' Certain commonly used
drugs increase parasympathetic function, pilacarpine being one,
aspirin, and most of the coal tar drugs being others.

The sympathetic, remember, is the ''braking system'' on the
vital functions of the animal and normally should not be in
greater tension than the parasympathetic, except during emer-
gencies requiring defense.

Hereditary dominance of, stimulation of, and overactivity of
the sympathetic division of the autonomic does the following
things:

Dilates pupils

Protrudes eye ball

Lessens lacrymation

Upper lid retracted, lagging

Intra-ocular hypertension

Lessens accommodation

Causes exophoria—reflex

Low adduction tendency

Inhibits ocular activities

Lessens mucous secretion of nose and throat

Lessens secretion and motility of stomach, stops digestion

Lessens salivary secretion

Slows peristaltic wave producing the common type of constipation

Increases pulse rate

Contracts arteries raising blood pressure

Increases blood sugar

Goose flesh and cold sweating hands, feet, under-arms

Increases body heat because:

    a. Increased chemical action

    b. Decreased radiation through the skin

Increases perspiration—palms, soles, under-arms

Diminishes quantity of urine

Contracts uterus

Activates:

    Thyroid

    Adrenal medulla

    Pituitary

    Gonads

    Muscles—these are not glands, but are involved in catabolic processes.

When the individual is in the vertical position "a column," the sympathetic tends to dominate the parasympathetic. This fact is a physiologic reason why the maximum physiological pupillary dilation that can be secured in the undrugged physiological individual is secured by requiring him to stand on his feet. This phenomenon is of value in the practice of ophthalmology and optometry when one is required to minutely inspect the fundus with the ophthalmoscope.

Incidentally, some authorities maintain that the extensor skeletal muscles are somewhat more under and susceptible to sympathetic stimulation than are the flexors, which they maintain are more under the parasympathetic control. The evidence here seems to be primarily circumstantial, in that the defensive muscles of the body, such as those used in combat or flight, are fundamentally the extensor muscles. Furthermore, the authorities are in agreement that there is a tendency toward exophoria when the pupils dilate in rage or fear. The lateral recti are classifiable as extensors. This mechanism may be one designed to give a wider field of vision at the expense of perfect fusion for the purposes of defense. Regardless of the apparent intention behind this connection, the work of *Eidelberg and Kestenbaum*[6] show an intimate connection between the act of convergence, mediated by the medial recti muscles, and pupillary contraction. These men have shown quite conclusively that pupillary contraction is not associated with the act of accommodation, but, *contra,* is associated only with the act of convergence, a medial recti function. Their experimental work lends support to the circumstantial conclusions mentioned above in that exophoria is a tendency to the loss of adequate convergence, and gives a physiological reason why exophoria should be accompanied by pupillary dilation as a result of a sympathetic stimulus.

The vagus, the tenth cranial nerve, is the most important nerve of the parasympathetic system; it is the largest of the cranial nerves and has the widest distribution as pointed out by *White.*[7]

It should be kept in mind that the autonomic nervous system control lies in the diencephalon and not in the cerebrum, however, the hypothalamus is to a certain extent under the control of the higher centers that lie in the forebrain including the corpus striatum and the dorsal portion of the thalamus. This control accounts for visceral responses associated with some sensory impulses such as ocular stimulation, from bad teeth, odors, pain, ocular pain, and by the act of seeing repulsive or horrifying scenes or actions within the visual field. *White* mentioned some of these sensory impulses specifically while others are a matter of common experience. Repulsive sights arouse emotions in individuals which take the form determined by their biotypes and by past experiences, which have set up conditioned responses in the several individuals. There is a direct connection betwen such emotional responses and associated visceral activities, and since the visceral activity is to a large extent controlled by smooth muscle, and smooth muscle is controlled by the autonomic, it becomes clear that the autonomic mediates these changes at the instigation

---

[6] Eidelberg and Kestenbaum, Jahrb. f. Physiat. v. Neurol., 46, 1, 1928.

[7] White, ''Autonomic Nervous System,'' p. 14, Macmillan, 1935.

of the thalamus and the hypothalamus as shown by *Clark*,[8] who also credits the hypothalamus in the central gray with being responsible for endocrine regulation, mediated by the pituitary, which causes the integration of impulses to the viscera and plays an important role in the control of the interior of the organism. The great importance of the pituitary in this connection will be discussed later.

Fridenberg[9] insists that by far the most numerous sensory stimuli which affect the vegetative nervous system by way of the diencephalon are those of ocular or visual origin. He further cites, in this connection, that the inter- and midbrain centers have to do with expression by way of the extraocular muscles. The lids, too, are intimately associated with sensory impulses from the eye as the sensory end-organ. He further implies that the eyes and ocular tissues may, due to their complicated structure, and the receipt of secretory stimuli, evolve or manufacture vitamins or even hormones as well. This latter thought is not new, having been voiced a number of years ago by another investigator whose trend of thought indicated he felt that possibly the choroid, the ultimate receptor of light energy after it passes through the retina, might be the source of endocrine production, or cause the production of a hormone-like secretion, which is distributed throughout the entire body by the blood stream after having been absorbed by the highly vascular bed of the choroid. These facts when considered in connection with the close relationship of the eye with the autonomic nervous system, due to its connections to both divisions and the more specific action of ocular structures, will be discussed in another place herein.

Sight should not be lost of the definitely reciprocal relationship existing between the two biochemical systems of the vegetative division. Biochemical here is used for the purpose of indicating that these reciprocal divisions activate different and antagonistic endocrine glands.

Carrying the thought of ocular effects further, there is the well known oculocardiac reflex, sometimes known as the *Emile Weil* sign, otherwise known as the *Aschner-Dagnini* reflex. This reflex

[8] Clark, "Structure and Autonomics," Brain, 1932, LV, 406.

[9] Fridenberg, "Endocrine Hormones and Vitamines in Relation to the Eye," Med. Rec., 148, 10, 378, Mar. 1938.

takes the form of a definite slowing of pulse rate, usually with an increased volume of the pulse pressure, resultant from mechanical stimulation of the eye ball.

It has been said above that some authorities include the eleventh cranial nerve as a component of the parasympathetic division of the autonomic. This inclusion is due to the fact that there are definite anastomoses between the spinal accessory and the cardio-pneumogastric—the vagus or tenth cranial nerve. The spinal accessory thus ramifies the head, neck, chest, and abdomen, as well as the trapezius and the sternomastoid muscle in the neck. In this way these parts are definitely linked with the act of respiration, circulation and digestion as accelerators, in addition to acting as restrainers to the heart muscle.

The diencephalonic nuclei coordinate all visceral activity, and all activity maintained through the function of smooth muscle, including those within the eyeball, as shown by *Papez*.[10]

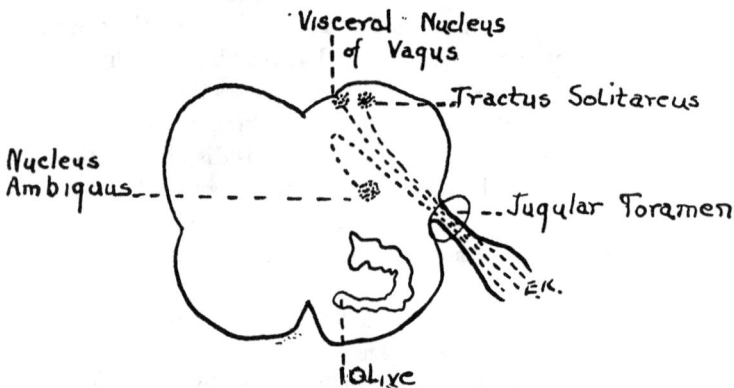

FIGURE IV

Grouped closely around the nuclei of origin of the vagus nerve in the diencephalon are nuclei controlling the vasomotor mechanism and the mechanism of respiration, the latter being usually spoken of as the "respiratory center." Incidentally, the cardiac center for inhibition in this same region.

White sets forth the autonomic innervation to the smooth muscle within the eye in the following summation:

[10] Papez, Arch. Neurol. and Psych., 1934, XXXII, 1.

Innervation of the Iris:
  a. Parasympathetic.

      Preganglionic neurones arise from cells in the Edinger-Westphal nucleus and run in the oculomotor nerve to the ciliary ganglion. Post-ganglionic neurones originate in the ciliary ganglion and run through the short ciliary nerves to the constrictor muscle of the iris.

      Function: Contraction of pupil and accommodation.

      Surgical application: None reported to date.

  b. Sympathetic.

      Preganglionic neurones originate from cells in the intermediolateral column, enter the highest two thoracic white rami, and ascend the cervical sympathetic chain to its superior cervical ganglion.

      Postganglionic neurones originate from cells in the above ganglion, ascend in the carotid plexus to the ophthalmic division of the fifth nerve, then run via the nasociliary nerves to the eyeball.

      Function: Dilation of pupil, protrusion of eyeball, and wideing of palpebral fissure.

      Surgical application: Resection of superior cervical sympathetic ganglion in cases of facial paralysis enables the patient to close his eyelids nearly completely.

It should be noted that both the sympathetic and the parasympathetic are here mentioned, the same being the two reciprocal divisions of the autonomic which we have heretofore shown to be under the control of the diencephalon. The ciliary body seems to be innervated only by the parasympathetic for contraction of that body. The ciliary ganglion, however, does receive sympathetic fibers and it would seem, therefore, that the mechanism governing the lessening of the accommodation, effected by relaxing the ciliary, may be brought about by intra-ganglionic inhibition. *White* also lists specific connections for autonomic innervation of the salivary glands, meningeal and cerebral arteries, the carotid sinus plexus, the heart, the pulmonary organs, the esophagus, upper abdomen, viscera, pelvic organs, the arteries, sweat glands, and erector pilae muscles, the latter having to do with the erection of hairs on the surface of the body. From this list it can be seen that the autonomic enervates all structures in the body containing smooth muscle, including the glands and smooth musculature of the heart and capillaries.

Due to action on the arteries and arterioles, the blood pressure within the arteries becomes a function of the autonomic, the sympathetic tending to raise the pressure by causing vasoconstriction, and the parasympathetic tending to lower blood pressure by producing vasodilation.

Kempf[11] observes that the pupillary changes during autonomic disturbance, particularly sympathetic over-activity, as a result of an emotion or stimulation of sympathetic nerve endings, result in active dilation of the pupil and cites the fact that man and all animals closely watch or observe the eyes of antagonists thereby learning the exact status of the defensive mechanism dominated by the sympathetic in such an antagonist. This "eye-to-eye" reaction in the animal kingdom being the surest and quickest way of interpreting the other's love reactions, safety reactions, rage reactions and combatting reactions. These reactions if forced to the extreme may result, as cited by *Kempf*, in compulsory evacuation of feces, urine and vomiting during fear or other intense stimulation of the sympathetic. Not only may fear of danger cause these reactions, but also fear of social criticism, financial losses, ostracism or a subjective sense of guilt on the part of the individual. *Chappell*[12] states that similar symptoms and disturbances of gastric and intestinal digestion may be caused by worry alone and that the application of known psychological laws of learning and forgetting, to enable the patient to eliminate worry, results in curing the physical condition.

In definite pathological conditions *Kempf* states that the sympathetic division of the autonomic integrates the circulation of the heart and striated muscle tone during subnormal temperatures of the body and during infectious diseases. This integration increases the work done within the body by striated muscle, increases combustion and raises body temperature to an optimum normal. If this process becomes too active there may be a fail-

[11] Kempf, Med. Rec., May 15, 1935, p. 478.

[12] Chappell, "Stomach Trouble Aided by Psychologists," S.N.L., April 14, 1934.

ure of the heat eliminating or dispersive system which may cause an actual rise of temperature in the body above the normal or even above safe limits in the form of a dangerous hyperpyrexia.

An interesting observation by *Kempf,* quoting *Orbel,* is that fatigued striated muscles, or contracted somatic muscles may be quickly restored to normal or above normal functional activity by stimulation of the sympathetic. This effect may be produced by temporarily inhibiting or paralyzing the parasympathetic by the use of such drugs as belladonna, or its alkaloid atropine, also by the injection of adrenalin.

Whitney[13] observes that a number of so-called diseased conditions are very persistent and nearly incurable because they have their origin in lack of syntony in the autonomic. Among such conditions are listed rheumatism, arthritis, asthma, angina pectoris, osteomalacia, poliomyelitis, Raynaud's and Burger's diseases, epilepsy, and high blood pressure. It is a matter of common observation that these diseases are practically incurable by the usual medical means, but that many of these may be favorably influenced by section or otherwise inhibiting the delivery of nerve impulses by either the sympathetic or parasympathetic. It seems that a continuation of irregular functioning of the autonomic results in a sufficient disturbance of the homeostasis of the body to cause irreversible structural changes. *Kern*[14] further supports *Whitney* in his statement to the effect that asthma is a result of a disturbed autonomic with resulting faulty endocrine secretion. *Crile* is a further supporter of *Whitney's* statement in reference to the contraction of blood vessels and also holds high blood pressure to be a result of autonomic disturbance. *Rivers*[15] shows quite conclusively that the mechanism for the development of arthritis and similar joint involvements is due to a lack of syntony in the autonomic, and further supports the statement and observations of *Whitney* above.

Cannon[16] made efforts to determine the mechanism whereby

[13] Whitney, "Accomplishments of the Physical Sciences," Science, 84, 2175, 216.

[14] Kern, "Autonomic and Endocrine System in Asthma," Med. Press, CXC, 5010, 37, 1935.

[15] Rivers, "Autonomic Diseases of the Rheumatic Syndrome," Dorsance, 1934.

[16] Cannon, "Autonomic Neuro-Effector System," Macmillan, 1937.

the autonomic effected responses. The result of this investigation indicates that smooth muscle under the influence of the sympathetic releases or secretes a chemical compound named "sympathin" by *Cannon* and *Bacq*. Sympathin, when absorbed into the blood stream and carried throughout the body, causes similar responses to those of adrenalin in many respects. It seems that adrenalin and sympathin act in a synergistic manner in that they aid each other in eliciting their respective specific responses. This action is demonstrable by the mere collection of a sample of sympathin, resulting from smooth muscle action under sympathetic stimulation, and injecting it into another animal in which the sympathetic is *not* under stimulation. The result of this injection, and the response of the injected animal thereto, proves it to be a new chemical compound to the economy of the animal into which it is injected. Since animals may live and do live under protected conditions, after complete ablation of the sympathetic, animals so prepared show the same characteristic responses under the injection of sympathin as do those who have intact sympathetic systems which are artificially stimulated in the laboratory.

Cannon also found that the impulses of the parasympathetic are mediated by a chemical effector which has since been found to be acetylcholine. Acetylcholine acts as a vasodilator in contrast to the vasoconstricting effect of sympathin. *Cattell*, of Cornell, used the mammalian eye to further investigate the effects of acetylcholine released by stimulation of the parasympathetic. Because the parasympathetic ganglion, the ciliary, can be readily removed he found that instillation of acetylcholine and other cholenergic substances causes pupillary contraction and some ciliary contraction. These experiments show that acetylcholine is effective in the eye when applied to the iris and ciliary body directly, in the absence of a ciliary ganglion. In order that there might be no residual nervous impulses reaching these structures during these experiments the post ganglionic nerve fibers were allowed to degenerate before instillation of the chemical effector of the parasympathetic-acetylcholine. These experiments would seem to indicate that many smooth muscles or smooth muscle cells which do not receive direct nerve fibers are caused to function by chemical means by the released acetylcholine in close proximity to them, within the same structure.

Obviously, prolonged sympathetic action, with its release of sympathin into general circulation, would result in general systemic changes, *contra* prolonged carrying of acetylcholine, of parasympathetic origin, by the blood stream would produce the opposite systemic changes.

Authorities are agreed that the response of the sympathetic is an *entire systemic integration* which takes place almost instantly following proper stimulus. This integration results in *all* of the effects listed under the action of the sympathetic heretofore; in other words this is a mass action, or an *undeviating pattern action* response. While a similar situation exists in reference to the parasympathetic, here we find a certain selective power whereby some functions may be more activated while others are not so much so. Some one has likened these responses, by analogy, to a piano in which the soft pedal, comparable to the sympathetic, damps all strings simultaneously, while the keyboard, likened to the parasympathetic, may have and usually does have some selective powers. It should be noted that once the strings are set in vibration by key action, operation of the soft pedal, whereas it damps all strings, has a greater damping effect upon those strings which are in vibration. Carrying this analogy into the autonomic nervous system we would expect to find that during the process of digestion with the stomach flushed with blood due to vasodilation, and the muscles in a state of activity, that a stimulation of the sympathetic would instantly, due to its antagonistic action, cause a cessation of the digestive acts, which at that time were in a state of activity, or in relation to other functions, over-active. This is exactly what happens, anger, fear, rage or any other over-activity of the sympathetic does stop digestion almost instantly in so far as it stops the functions necessary for the maintenance of the digestive act. A similar situation exists in almost any other parasympathetically controlled function.

Another, and to some extent more dangerous to the individual, phenomenon is that *conditioned responses*, which are artificially developed association tracts, are instantly eradicated, temporarily or permanently, as a result of sympathetic stimulation. Conditioned or learned responses required cerebration during the process of their acquisition, but once acquired they become some-

what like reflexes in that they no longer require conscious effort
to cause them to go into action. Practically all of the associated
functions of vision fall in this class, i.e., those functions operated
by the extra-ocular muscles and their associated nerves and nerve
centers. These associated ocular functions temporarily disappear, or are considerably lessened in their power for continued
function, following stimulation of the sympathetic, thus destroy-
ing wholly or in part the functions required for accurate convergence, stereopsis, depth perception, and accurate fixation
binocularly for the maintenance of single simultaneous binocular
vision. Disturbance of these functions may result in out of focus
images and blurred vision, or in an extreme case, might result in
diplopia—double vision.

Henderson,[17] quoting Yale's scientific magazine, mentions a
possible danger to life due to the nullifying of conditioned responses under sympathetic stimulation. He cites that the art of
driving an automobile, without undue fatigue, is largely one of
the development of conditioned or learned reactions. A driver,
driving at a rapid rate and suddenly confronted by a situation
such as an impending collision or a violent skid becomes integrated defensively, thus losing his conditioned responses temporarily and "jumps." By jumping is meant a sudden pressure
of both feet upon a support which is a normal and natural reflex
phenomenon in the presence of sudden danger. If one of his
feet is on the floor and one on the accelerator during the action
of this natural response, his statement to an officer requesting
information as to how the accident happened, will be, "Why,
officer, I don't know. The car just went out of control." Obviously, the car did not go out of control, but the *driver did.* He
went out of control because his conditioned response, for the
purpose of stopping the car, is to lift the right foot, move it to
the left, place it upon the brake pedal, and then press down.
This being a conditioned reaction disappeared under sympathetic
stimulation, because it is not a normal reflex to *lift* one's feet in
the presence of danger; the natural response being to press down
with both feet. Not only does a driver faced with the situation
mentioned press down with both feet, but he strongly seizes the
steering wheel and "freezes" it usually in the position in which

[17] Henderson, "Driver Lost Control," Yale Scientific Mag.

it was at the instant of sympathetic stimulation, which might have been in a straight ahead position or at some angular position. If the wheel was in some angular position, the car would probably leave the road, mayhap head into a ditch. Here the natural righting reflex would come into effect, and cause him to turn the wheel farther in the direction in which it was turned, causing the car to turn over as a result of the sum of gravitation and centrifugal forces, despite the fact that the *learned* or conditioned response is to turn the wheel in the same direction that the rear of the car is sliding and thus stop the skid and prevent turning over. As a result of sympathetic stimulation the driver has lost control of himself and the car obeying his faulty control, and the natural laws of gravitation and centrifugal force does the rest. Strictly speaking, of course, the driver in such a situation is not to blame, because the reactions which take place within his body are inherent therein, just as they are inherent in all animal bodies.

Just as an autonomic nervous system in syntony is an aid to healthful living and a ripe old age, *contra,* an autonomic out of "syntony" results in ill health and a short life, either due to disturbed function or forceful dangerous contact with environment.

Hereditary dominancy of the parasympathetic, which is incidentally under the control of and dominated by the central gray, builds and constructs a far different structural body than is the case of the hereditarily dominant sympathetic. The result is two widely different biotypes, which will be discussed at length hereinafter.

The foregoing might well be summarized into the two following statements: The parasympathetic keeps the animal alive operating his vital mechanism; the sympathetic takes care of him in danger by activating his defensive reactions.

Obviously, neither of these two divisions should dominate the other normally and should only act as a check upon each other. In the presence of an equalized checking tension between these two divisions we have the *syntonic* state, or as *Raup*[18] calls it, "equilibrium" or "complacency."

[18] Raup, "Complacency, the Foundation of Human Behavior," Macmillan, 1926.

Figure V is by way of an analog diagram to the situation that exists in the body. The opposite ends of the teeter labeled "P" and "S" represent the parasympathetic and the sympathetic divisions respectively of the autonomic in a state of syntony as indicated by the solid lines. The teeter in a state of imbalance, as indicated by the broken lines, represent the sympathetic as being in a greater state of tension or dominant over the parasympathetic. If this were a real teeter-totter it would be possible for a third individual to stand astride the center and by throwing his weight more upon one foot than the other he could restore a state of equibrium as indicated by the solid lines. This individual in the diagram has his head labeled "central gray,"

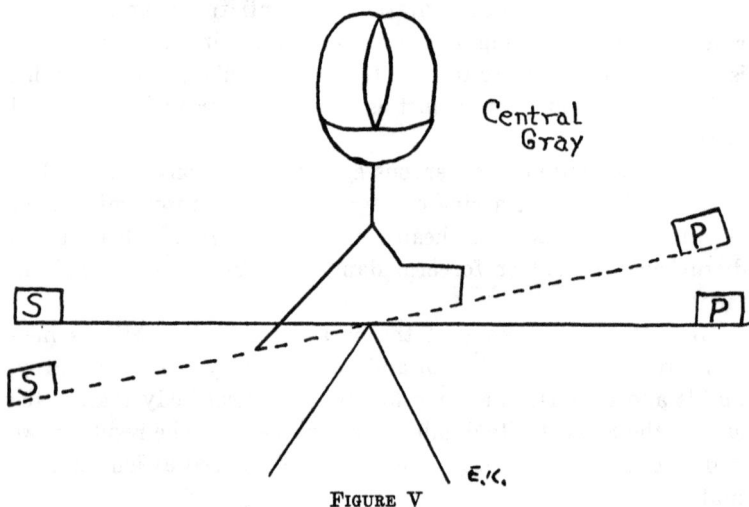

Central Gray

S

P

S

P

E.'.'.

FIGURE V

of which the mass of the thalamus is the most important group of nerve cells from which eminate efferent impulses upon the receipt of afferent impulses. Stimulating or depressing the thalamus alters the state of equilibrium existing between the two divisions as represented by the analog. We shall later show that it is possible by using purely optical means, and the resultant nerve impulses which pass over the shortest, most direct and most highly organized nerve pathways, those from the eye, to alter this state of tension in the central gray in such manner as to disturb equilibrium, or to restore equilibrium within physiologic limits once it has been lost. Equilibrium here is used in the sense of syntony, an equalized or like tension between these two divisions.

REFERENCES

Oaks, ''Physiologic Hypothesis of Allergy and its Effective Treatment,'' Eye, Ear, Nose and Throat Monthly, 1937.

Jones and Goyert, ''Treatment of Vegetation Excitation,'' Clin. Med. Surg., p. 498, *et seq.*, Oct. 1936.

Blackmar, ''Symposium on Nerve Clinical Approaches to the Autonomic Level,'' Laryngoscope, Dec. 1934.

Morris, ''The Autonomic Nervous System,'' Ind. Optometrist, May 1933.

Brock, ''Basis of Clinical Neurology,'' Pray *et seq.*, Wm. Wood and Co., 1937.

Pottenger, ''Symptoms of Visceral Disease,'' p. 105 *et seq.*, C. V. Mosby Co., 1922.

White, ''Autonomic Nervous System,'' page 17, *et seq.*, Macmillan, 1935.

Cannon and Rosenblueth, ''Autonomic Neuro-Effector Systems,'' Macmillan, 1937.

Chapter VIII

BIOTYPES

That the physical make-up and structure of the individual governs his expression of function is conceded by authorities in medicine, anthropology, psychology, physiology and psychiatry. The study of the structure in this relationship considerably antedates the Grecian philosophers, one of whom stated that "the greatest study of mankind is man," but who apparently did little more about it. An early reference to structure in relation to response and as it relates to temperament and environment is found in the thirty-first chapter of the book of Proverbs in the Old Testament wherein is set forth an ancient Hebraic description of an ideal woman.

At a much later time Hippocrates, the "father of medicine," described four constitutional types, the study of which he held to be of vital importance to the followers of physic. His contention was that a knowledge of these constitutional types, now known as biotypes, was important because the several types were prone to suffer from ailments usually not found in the other biotypes. *Lombroso*, of the University of Padua, undertook at one time the description of what he thought was the criminal type. His descriptions are replete with minutae, such as pointed lobes on the ears, ear lobes attached to the cheek at the ramus of the jaw, position of eyes, and the bony formation of the face and head generally. Efforts to use *Lombroso's* criteria fell far short of a practical application when applied to large numbers of known criminals or to non-criminal civilians.

The failure of the *Lombroso* criteria resulted in a more scientific approach to classification of human biotypes. It was soon learned that certain gross structural formations apparently do bear a definite relationship to and an association with the temperament and the personality of the individual. Many terms have been devised which are in part descriptive of these biotypes. These descriptive terms have one thing in common, in that they describe the relative width of the head and body in their relation to the vertical height of the face and total body height. The following tabulation indicates some of the terms devised by a number of investigators:

A. de Haller, 1750			Type athletique	
Halle, 1797— —Francais—	Type thoracique	Type abdominal	Type musculaire	Type nerveaux et cephalique
Cabanis, 1802— Francais—			Type musculaire	Type nerveux
F. Thomas de Trois-vevre, 1821 Francais—	Type thoracique	Type abdominal		Type cranien
Tostan, 1826— Francais—	Type respiratoire	Type digestif	Type musculaire	Type cerebral
De Giovanni, 1877— Italien—	1 combination	3 combination	2 combination	
Benecke, 1878— Alleman	2 type	normal	1 type	
Virenius, 1904— Russe—	Type epithelial	Type connectif	Type musculaire	Type nerveux
Claud Sigaud, 1908— Francais—	Type respiratoire	Type digestif	Type musculaire	Type cerebral
Kretschmer, —Allemand con temporain—	Type asthenique	Type picnique	Type athletique	
Classification des veterinaires zootechniciens	Type faible sec de laiterie	Type faible et gonfle deau —de boucherie	Type fort de travail normal	Type infantile
D'apres les principes pathologiques classification— Russe?—	Type asthenique	Type apoplectique	Type normal	
Bryant, 1913— American—	Type carnivore	Type herbivore	Type mesoplastique	Type subplastique—?
Bounak,— Russe—contemporain—	Type stenoplastique	Type euryplastique		
Bean, 1912— American—	Type hyperontomorph	Type mesontomorph		

It seems that a majority of tabulations incorporate these three fundamental groups: 1. Those with narrow heads and faces. 2. Those with relatively wide heads. 3. Those with relatively square faces and heads. The descriptive terms which will be used hereafter to describe these biotypes will be *asthenic,* for the narrow head and slender bodied individual whether short or tall; *syntonic* for the square head and face with a relatively athletic type body, of medium height; and *pyknic* for the wide head, wide bodied individual whether short or tall. It must here be pointed out that there also exists a difference in personality and temperamental responses in the three types which is to some extent correlated with height. This fact, however, will generally be ignored, except for mention of specific deviations or characteristics which have definitely been correlated with height. *McCurdy*[1] states that generally speaking the male tends toward the asthenic, while the female tends more toward the pyknic biotype. *Leo Loeb,*[2] of Washington University Medical School, states that the response of individuals of the several biotypes extends even to the responses of their component cells and tissues, indicating that cells themselves could be classified as asthenic, syntonic or pyknic types, if proper criteria were available for their microscopic study. *Haggard and Fry*[3] state that the temperamental qualities of each of the biotypes is a function of and transmitted as a hereditary quality. The intellectual and the emotional equipment of the individual are inseparable. *Haggard* insists that intellect is hereditary. This is a logical conclusion which may be drawn from the findings of *Lewis,*[4] that intellect and constitutional biotypes have a definite association.

Haggard further states that the behavior of the human being is inborn in the individual along with his physical form and physique, and that these basic traits of personality, while they differ in each individual, are fixed and permanent throughout life, and that it is these fixed qualities which dominate his behavior.

The character of the individual should definitely be differen-

[1] McCurdy, Med. Rec., Sept. 15, 1937.

[2] Loeb, Leo, "Biological Basis of Individuality," Science, 86, 22, 18, 1–5.

[3] Haggard and Fry, "Anatomy of Personality," Harper Bros. 1936, p. 46.

[4] Lewis, "Borderline of Sanity," Knowledge.

tiated from his personality and temperament. The latter two are inherent hereditary qualities, while character is the reaction of an individual's fixed hereditary qualities to the environment in which he is reared. The combination determines the pattern of reaction and the modes of behavior which we recognize as the character. Each individual, therefore, reacts to his environment as only he can react to it. Structural analysis by anthropological measurements permits the differentiation of these inherent qualities, knowing which, it is within the limits of probability to predict with a high degree of accuracy the response of an individual to any reasonable environmental change or condition. *Haggard* implies that there is a body of students who scoff at the idea that physique governs temperament and responses. He cites them, however, to the stubborn fact that it is easy to demonstrate that a relationship does exist. For instance it is almost impossible to find an extreme pyknic among those who are outstanding or who have achieved fame as philosophers. Pyknics are too practical minded to become philosophers. *Kretchmer*[5] in his "Psychology of Emotion" lists the following characteristics of the two types:

TYPES OF CONSTITUTIONS AMONG THE HIGHLY GIFTED

	Pyknic	*Asthenic*
Literary men	Realists	Romanticists, writers of extreme pathos, formalists
Scientists	Humorists Impiricists describing things just as they appear	Exact logicians Systematists Metaphysicians
Leaders	Tough, pushful men; happy spirited organizers. Judicious and understanding mediators	Pure idealists; despots and fanatics; cold, calculating men.

Temperament and personality being definitely associated with the structure of the individual as they are, and since the temperament, personality and responses of the individual are known to be controlled by the thalamus, as cited by *Haggard, Cannon, Kempf,* and others, leads us to one and *only one* conclusion, i.e., that the structural development of the individual is basically a

[5] Kretchmer, "Psychology of Men and Geniuses," Harcourt Book Co., 1931.

function of the thalamus. The exact part the thalamus plays in
conjunction with the chromosomes, involved in the transmission
of hereditary qualities, is somewhat a matter of conjecture. It
is a well recognized fact, as heretofore stated, that the thalamus
and the autonomic nervous system exercise response control over
the endocrine glands. The endocrine glands operating individually or severally play an important part in the growth and
length of the long bones, the shape and formation of the irregular
bones, and control their *rate* of growth. Hence it does not seem
to be a strained conclusion that the thalamus and perhaps the
corpora quadrigemina constitute the basic control over structure
within the individual.

The foregoing conclusion, while at first somewhat startling, is
susceptible of a reasonable demonstration. Ablation of the cerebrum, removal of parts of it surgically, or congenital absence
thereof, interferes with and prevents intelligent thought. The
individual so deficient, however, would still possess his impulses
and desires. He would still live and grow, he would exhibit emotion, in other words he would possess a personality and a temperament. He would, though, lose all power of restraint and
would be incapable of developing what we understand as character, nor could he be caused to fit into a civilized social order. He
would be a brute. He would, therefore, react instinctively only,
because instinctive actions are unyielding modes of behavior
which are inherent in the structure of the nervous system. The
only way these instinctive actions could be altered would be to
change the structure of the animal and his nervous system.

But the structure of the body, the color and type of hair, and
the color of the eyes are generally recognized as hereditary qualities and nothing can be done following the instant of conception
to prevent their development and occurrence as dictated by the
hereditary. No change in environment, no type of training, no
form of education will change a tall slender, blue-eyed blonde
into a short, stocky brunette. They remain as their heredity
determines they shall. Each of these two extreme biotypes, however, may be developed along certain character lines which are
not wholly antagonistic to their structure. Nor can their
quotient of intelligence be improved beyond their hereditary
acquisition by any of these things, to the confusion and chagrin

of doting parents, educators, child psychologists, and perhaps the worst of all, the social reformers. Efforts to make such changes are beyond the capabilities fixed by the structure of the individual and are mere wishful thinking, and should have no place in science. This may be a "bitter pill" and may hurt the pride of many parents, but the facts must be faced and the sooner educators, reformers and parents begin to conduct their activities in line with known facts the sooner real constructive work can be done. *Stockard,*[6] of Cornell Medical School, for a number of years has conducted a long series of animal breeding experiments principally with dogs. Simultaneous checks of reaction patterns of dogs obtained by various breeding and inbreeding experiments have shown that these patterns are as diverse as the structure of the dogs themselves, yet conform to the structural setup of dogs. Also, that the reaction patterns are anticipatable by one who has made a prior study of these animal structures in connection with their reactions. All that is required is a close observation of the animal. It should be noted, though, that all of these dogs possess the basic canine temperaments. Even so, the reaction patterns within the canine temperament are almost as numerous as those found in the human race. *Stockard* even goes so far as to picture in parallel columns on the frontispiece of his book "Physical Basis of Personality," dog biotypes which bear remarkable resemblances to the there pictured human biotypes.

Davenport[7] in his monograph, "Biology of the Individual." lists a number of diseased conditions which occur in the several biotypes, associating with the asthenic a high incidence of influenza, colds, tuberculosis, melancholy, thyroid involvements; with the pyknic stomach troubles, dropsy, apoplexy, kidney diseases and cancer. He also states that heart trouble, arteriosclerosis and diabetes are especially common in the latter biotype. A more complete list of diseases commonly associated with the extreme biotypes will be given later.

Since in the further development of this thesis ocular problems will be discussed, it seems well here to list ocular structural background. *Mohr*[8] lists the following: microphalamus, coloboma, in

6 Stockard, "Physical Basis of Personality," Morton, 1931.

7 Davenport, "Body Build and Inheritance," Bio. of Individ., Vol. XIV.

8 Mohr, "Heredity and Disease," p. 126, *et seq.*, Norton, 1934.

some cases cataract, lens luxation and lens subluxation, anaridia, extreme myosis, degeneration of the macula, retinitis pigmentosia, a form of hereditary optic atrophy known as Lieber's disease which affects males, but is transmitted by the female in like manner as hemophilia. *Mohr* also lists nightblindness, red-green blindness, some cases of glaucoma, hydrophthalmus, commonly called bull-eyes, albinism, infantile amaurosis, Huntington's chorea, nystagmus and allergic hypersensitivity of the conjunctiva. Here is an imposing list of ocular departures from the normal, most of which are incurable except, perhaps for some relief in cases of Lieber's disease and nightblindness as will be mentioned later. It should be noted that the majority of these deviations are structural anomalies rather than functional.

Travis[9] states that seventy-five percentum of stutterers are asthenics, but that about ten percentum are of the syntonic or athletic type build, but that *not one* stutterer was found in the pyknic type. This would indicate that there is either an emotional instability or a functional anomaly of the nervous system, both of which are predicated upon structure, which is hereditary, indicating that perhaps this grave and embarrassing difficulty has a hereditary basis. In this connection authorities apparently agree that the emotional instability inherent in the nervous structure is such that contact with the environment results in cross-purpose responses resulting in the stuttering.

Sinnott[10] insists that the form of the structure is an outward visible expression, fixed in material shape, by an inner organized equilibrium. The writer would qualify this statement by saying that the *extreme* asthenic and the *extreme* pyknic are more likely to be the result of a lack of an organized *equilibrium* in the individual. Since we can only examine the end product of the nervously controlled development and cannot examine the process whereby this development takes place, we are forced to rely upon the observable anthropological structure as the criterion for anticipating and prognosticating the type, kind and degree of function. In other words, we study the inner mechanism by observing the outward form it produces.

Todd[11] lists the following factors which should be studied in

9 Travis, "Stutterers Usually Asthenics," S.N.L., Sept. 23, 1933.
10 Sinnott, "Morphology a Dynamic Science," Science, 85, 2194, 61–65.
11 Todd, "Physician as Anthropologist," Science, 83, 2164, 590.

connection with the determination of responses of the several biotypes: Growth increments and proportions, physical maturity progress, weight gain, and the analysis thereof, dento-facial development, brain potential, and muscle action currents, mental expansion, psycho-motor development, hand-eye coordination, motor development, acquired skills, manual steadiness, and lack of tremor, dexterity, emotional stability, intellectual and social adjustment, interests and talents, personality ratings, vocational leanings, and choice. Obviously, such a study will bring to bear upon the problem of the biotype and its responses, numerous facts which once correlated will give reliable criteria for study of the biotypes as they present themselves. It must be kept in mind in this connection that the acquisition of skills, emotional and social adjustment and other acquired characteristics do not obliterate the genetic biotype of the individual, nor do they alter it beyond the limits of capabilities inherent within the physiological limits of the biotype. *Draper*,[12] of Columbia, says that the capacity of an individual to react or not to react to an external stimulus is a constitutional quality which is specific in terms of body size and formation and applies to function as well as to the psychic pattern. *Draper*[13] specifically mentions the following diseases which are characteristic of certain facial patterned biotypes: Nephritis, a face the sides of which are parallel, narrow, and with a relatively wide chin; tuberculosis, a slightly narrow face, the sides slightly converging toward a relatively pointed chin; gastric ulcers, a face wide at the top tapering down across the cheeks to the chin, the chin itself being fairly square. In this latter connection he shows photographs of two male patients both of whom suffered from gastric ulcer and whose facial resemblance is strikingly like that of identical twins; gall bladder disease, a head wide at the top and across the zygoma ovally slanting toward the chin; pernicious anemia, a wide parallel sided face with a square jaw and a wide chin. Carrying his identification of these disease tendencies further, *Draper* even finds a striking agreement of plaster casts made of the upper and lower teeth in these patients. Continuing, *Draper* again confirms findings of others to the effect that intellect is an inherent quality

[12] Draper, ''Human Constitution,'' p. 222, Saunders, 1924.
[13] Draper, ''Disease and the Man,'' Macmillan, 1930.

and bears no relationship to environment or early training. He
does agree, too, that the so-called feeling states, emotional and
functional responses of the individual may be altered or varied
by the impact of physical agencies, chemical or bacterial agencies
or by emotional stresses constantly present in the environment,
the latter being intimately associated with the sympathetic endocrine involuntary mechanisms. *Freeman* is quoted in ''Bloodless
Phlebotomy'' on the result of 1400 autopsies of patients who had
died in an institution with which he was associated. He found
that certain diseases common to one biotype were never found in
the other biotypes. *Freeman* found that his autopsies served to
confirm the deductions of other workers in his field, who had
studied functional responses and anomalies in their association
with structure and who had by their study arrived at the conclusion that ''structure governs function.''

In discussing the constancy of the structural biotype, *Boaz*[14]
made a study of immigrants and reached the conclusion that no
environmental change to which the immigrant was subjected here,
upon coming to America, was sufficient to change in any manner
the biotype.

A significant finding by *Kempf*[15] is to the effect that the facial
nerve proper, and of the nervus intermedius of Wrisherig, both
being connected in the visceral area of the medulla along with
the nucleus of the vagus, accounts for the gradual adaptation of
the facial muscles and facial expression attendant upon nervous
response of the individual biotype resulting from his contact
with his environment. It is these connections which account for
many facial and expressional modifying signs of the biotype,
which may be observed and interpreted by one trained in their
significance. Since these are a result of environmental stresses
and response thereto, they serve to enable the observer to read
the environmental past of the individual in terms of his structure, as a structure-compelled response thereto.

Gonzales is quoted as having discovered and proved that the
two sides of the face of any individual, structural and in terms
of muscle tension, represent two different phases of the personal-

14 Boaz, ''Effect of Environment Upon Immigrants,'' Science, 84, 2189,
523.

15 Kempf, ''Physiology of Attitude,'' Med. Rec., May 15, 1935.

ity, the right side with its bone and muscular development is a criterion of the continued conscious reactions of the individual, while the left side reflects his unconscious and emotional life. *Ashton-Wolfe*, formerly of the French Surette, reported that at one time a criminal gang in Paris had made use of the findings of *Gonzales* for their own ends. The method was to examine the windows of exhibits by photographers for full front views of the photographer's patrons. Upon finding a face in which the left side showed an infantile emotional status, their method was to learn the name of the patron whose photograph was on exhibition and to make efforts to solicit that person to participate in their various schemes and crimes. These facts came to light following the arrest of the chiefs of the gangs. In this connection, there have been published on several occasions composite photographs of men in public life consisting of three front views, one made of the two right halves, in the center the normal face as a whole, and the third picture made of two left halves. It is not uncommon following this procedure to find that the face made of the two right halves possesses an innocent lamb-like expression, while that made of the two left halves is quite often brutish and repellent to the observer.

A recent investigation of the biotypes associated with crime, criminal tendencies, and insanity by *Hooton*,[16] published by Harvard University Press, lists the types of crime usually committed by the following nine structural biotypes:

1. Short, slender individuals with narrow faces.
2. Short, medium build with square faces.
3. Short, stocky build with wide heads.
4. Medium height, slender with narrow faces and narrow heads.
5. Medium height, with wide faces and heads.
6. Tall, slender with narrow faces and heads.
7. Tall, with square faces and heads.
8. Tall, with wide faces and heads.
9. Tall, stocky build with wide face and forehead.

It is interesting to note that the types of crimes committed by criminals falling in one of these nine classes differ quite radically from those in the other classes, although as is to be expected there

---

[16] Hooton, ''Crime and the Man,'' Harvard Press, 1939.

is some overlapping in the degree of crime. A further interesting observation is that the crimes of violence such as burglary, first and second degree murder, are more specifically associated with the wide headed groups, while crimes requiring finesse such as fraud, forgery, and similar types of crime are usually committed by those with the narrow heads. *Hooton* analyzed the anthropological measurements of practically fifteen thousand criminals and a similar group of non-criminal civilians and reached the conclusion that the criteria are sufficiently different in the two groups, by rather large percentage, enough to permit anticipatory diagnosis of criminal tendencies in an individual. In his conclusion as a result of over twelve year's work in this field he makes the point that scientists and criminologists have consistently refused to learn a lesson from the organic structure of the individual and that his behavior is a function of that structure, and that the only possible way to improve behavior upon any permanent basis is to prevent reproduction of the structures which are found to be associated with the tendencies discovered by his investigations. He classes as imbecilic the assumption that culture, training and education will in any way prevent anti-social behavior manifestations which are inherent in the structure. He insists that the positive anthropological differences between the various classes of criminals and the deviation between criminals and civilians is great enough to enable the early segregation of those shown as possessing the physical structural criteria of the criminal.

Continuing further, *Hooton* says that among rapists there are quite a few full bodied ruffians, but by far the majority are ''shriveled runts.'' Thieves and burglars are constitutional inferiors who are physically stunted or mal-nourished or both. Robbers incline to be wiry, narrow bodied and tough but not notably undersized. That forgers and criminals involved in fraud show the least variation from the civilian. Thugs tend to be of the bulky type. He insists that the offense groups above show head and body measurements that sharply differentiate them from the generalized type of physique. The tall, thin men tend toward calculated murder and robbery, tall heavy men tend to kill or are associated with forgery and fraud; the thin undersized men tend to steal or burglarize; short heavy men assault, rape and other sex crimes. While men of rather mediocre body build tend to break the law in unconstitutional manners. The conclusion he reaches is that those of extreme bodily build incline to specialize in the types of crime they commit, impelled by their constitutional structure.

A further yet interesting conclusion reached by *Hooton* is that the structural elements which make for anti-social acts differ from that of those who have a disposition toward the mental diseases, particularly schizophrenia. He deplores our present ignorance of human biology in terms of anatomical and structural criteria of organic inferiority, which might well be used to determine, well in advance, those of low mentality or a worthless criminally inclined individual. He directs psychiatrists to a consideration of the physical organism of the individual and the individual physique as a means of aid in determining the type of mental diseases which may co-exist in the patient. He takes the medical profession and other biological professions to task for ignoring for centuries the obvious relationship between structure and function. He cites medicine, particularly, as having turned away from man himself down into the blind alleys of a study of diseases, bacteria, pharmacology, hygiene and immunology. He reaches the further conclusion that the quantitative and qualitative structural variations are the source of human behavior both for good and evil, and that these variations are definitely and hereditarily inherent in the individual, due to the biotype.

It seems that a tabulation of biotypic variants between the pyknic and the asthenic might well here be listed. These variants, as will be noted, consider the entire structure and hirsute adornment of the two extreme types. It should also be noted that in these classifications the type and kind of food which should predominate in the diet is also listed:

ANIMAL AND HUMAN CLASSIFICATION

Asthenic—carnivorous:
    a. Whole figure, including skeleton is light.
    b. Slender.
    c. Skin soft and delicate.
    d. Hair abundant on usual places, may be found in unusual locations.

e. Tall and slender, OR
f. Small and delicate.
g. Head may be proportionately large, BUT
h. Face and jaw narrow.
i. Ears large and prominent, projecting outward and forward.
j. Torso longer and narrower than so-called normal.
k. Lumbar spine flexible, make fancy dancer, hurdler, acrobat.
l. Thorax fair size.
m. Lungs and heart small under x-ray.
n. Stomach long and tubular instead of pear-shaped.
o. Stomach attachments not firm, prolapsus.
p. IMPORTANT: intestine ten to fifteen feet compared with a normal twenty.
q. Muscle fibers long and slender, quick action, but NOT sustained action.
r. Intestinal walls thin, lumen small.
s. Arms and legs slender, see q.
t. Slender feet, HIGH arches.
u. Hands and fingers long, slender tapering.
v. If accumulate fat always soft, lost quickly and indicates poor health.

Pyknic—herbivorous:
a. Body on heavier lines throughout.
b. Skeleton heavy.
c. Muscles large with coarse fibers.
d. Skin coarse.
e. Scanty hair, lost early.
f. Excess of fat throughout body.
g. Flesh hard and firm.
h. Head round.
i. Face broad.
j. Neck short and thick.
k. Jaw square or heavily round.
l. Ears usually flat, small.
m. Chest massive, antero-posterior and lateral.
n. Shoulders broad and massive.
o. Body broad and relatively short.

p. Abdomen, broad and deep.

q. Stomach large and pear-shaped.

r. Small intestine long, twenty-five to thirty-nine feet.

s. Large intestine five to eight feet.

t. Joints lack flexibility.

u. Legs large.

v. Knees straight.

w. Feet broad compared with length.

x. Arms heavy and attached back on shoulder.

y. Hands broad and "chubby."

A number of years ago the writer set up criteria for those two biotypes as a result of his investigations prior to a knowledge of *Kretchmer*, author of the above classification, and the writings of *Stockard*. The writer's classification follows:

FACIAL AND BODILY SIGNS AND CHARACTERISTICS

Asthenic	Syntonic	Pyknic
Thin triangular face	SQUARE FACE	Full round face
Thin upper lip—as rule		Full lips
Long nose, high bridge		Small depressed nose
Narrow bridge		Wide bridge
Rapid pulse		Slow pulse
Hollow cheeks		Full round cheeks
Mouth closed, eyes open		Mouth open, eyes closed
Pointed chin		Globular chin
Long neck		Short neck
Long extremities		Short extremities
Bass voice		Tenor voice
Trunk short and narrow		Trunk long and full
Shoulders, square, high, angular		Shoulders sloping
Crowded illy-set teeth		Teeth, even, not crowded
High cheek bones		Depressed cheek bones
Bony		Fleshy
Pale		Red
Tall—usually		Stodgy
Lips pale		Lips red to purple
Eyes large, maybe narrow P.D.		Eyes small—wide P.D.
Delicate texture skin		Rather coarse skin
Narrow head		Wide head
Tend to become fleshier after 35		

A list of certain reaction characteristics of these extreme biotypes has been accumulated by the writer over a period of years from the published books and papers, too numerous to bibliograph herein, by the following: *Tridon, Draper, Stockard, Daniel, Stratton, Good, Richmond, Freeman,* and others.

A search of medical literature resulted in the following list of diseases, abnormalities and functional tendencies which tend to occur in the extreme biotypes as a result of their structure:

FUNCTIONAL TENDENCIES

Asthenic	*Syntonic·*	*Pyknic*
High metabolic rate		Low metabolic rate
Hyperopia		Myopia
ESOPHORIA		EXOPHORIA
Dyspepsia		
Hypotension		Hypertension
Hyperthyroid		Hypothyroid
Headache		Apoplexy
Melancholia		Fatty degeneration:
General debility		Heart
Wasting diseases		Kidneys
Dizziness		Inflammations—gouty
Intestinal cramps		Rheumatism
Heart failure, Class IV		Scrofula
Menstrual cramps—at times		Tumors
Gastric ulcers		Menorrhagia
Tumors—cystic		Asthma
Acidosis		Gall bladder
		Diabetes—Mendelian
		recessive
		Alkalosis

These conditions are more severe if any of the above conditions cross the biotype or column.

Chapter IX

THE ENDOCRINES

In a previous chapter wherein the autonomic nervous system was discussed both in relationship to the control of function directly, and in terms of the development of an individual's temperament, it was stated that a part of this mechanism involves the control of the endocrine system by the autonomic. A further study of the endocrine system in this relationship, and the interrelationship of the glands themselves, is necessary in order to understand some of the conclusions which are later to be drawn from the facts.

It has become somewhat axiomatic in the medical and biological arts and sciences to suspect a disguised or veiled endocrinopathy in all cases where the symptoms are not definitely attributable to demonstrable physical or physiological findings. Restated, in the professional practice of these arts, this means that in all cases which do not respond to the ordinary or usual procedures, endocrine involvement is suspected.

The general make-up and the biotype of the individual has been shown to be a function of the central gray in the brain acting through the autonomic. Such action must include the association of the autonomic and the endocrine glands and their effect in producing the original structure, even including the modifications of the structure which may take place in later life. Endocrinologists also make use of the relative size and shapes of teeth in the upper jaw, the front six teeth, in determining the early developmental status of the endocrine system. The assignment of teeth to specific glands is as follows: The central incisors by their relative size and mass are a clue to the early developmental status and activity of the hypophysis, otherwise known as the pituitary gland. Usually there is no differentiation in this assignment in terms of the two major divisions of the pituitary, although there is some thought that it is the anterior lobe of the pituitary which should be definitely meant when using these two teeth as a criterion. The lateral incisors are assigned to the gonads, not in reference to their specific reproductive function, but to their interstitial or hormonal secretions alone. The canines, both by their relative size and mass in terms of the other four of the

101

front teeth in the upper jaw, are a clue to the early developmental status and degree of activity of the adrenal glands, presumptively, the adrenal medulla.

While these teeth signs give some clue as to the developmental size and perhaps the quality of the secretion of these glands, they do not enable the practitioner of the biological arts to determine their present status in later life. By "present status" is meant an over- or under-secretion of the gland at the time of the examination. It follows, then, that the criteria for the determination of either hypo- or hyper-secretion at a given time are a necessary bit of knowledge for the practitioner of these arts.

Recent investigation indicates that the major activities of all the endocrines are controlled by the centers in the diencephalon, particularly that portion known as the hypothalamus, *Goldzieher*.[1] The nerve centers are inter-related to the endocrine glands, but conversely these glands also exert certain specific influences upon the nervous system, constituting a mutually interacting and co-acting system.

Since some of the endocrine glands are activated by the sympathetic division of the autonomic, while others are activated by the parasympathetic division, it should be understood that stimulation of, dominance of, or over-activity of either of these divisions automatically induces a reverse tension or activity in the opposing autonomic division. This appears to be the mechanism whereby equilibrium of function is maintained, i.e., causing the body to stay within the limits of the normal physiological deviations.

The proper approach to the study of this mechanism appears to be a brief, yet detailed, study of the several glands and their normal functions. This study will be followed by some of the physiological effects of deviations from the normal with the establishment of criteria by means of which abnormal deviations can be recognized. An observed phenomenon in relation to the endocrine system is that not only are these glands activated by the autonomic at the instigation of the diencephalon, but the secretions of the glands themselves, which are carried in the circulation by the blood stream, seem to be necessary for the proper activation of synergistic glands or the proper inhibition of antagonistic glands. This is just another way of saying that the

[1] Goldzieher, "Practical Endocrinology," Appleton-Century, 1935.

function of each of the endocrine glands is not only dependent upon the autonomic, but is dependent upon the other glands of the endocrine system.

Since it is generally conceded that the pituitary is the "chairman of the board" of the endocrines, it will be first considered here. The pituitary is not immediately necessary for life. Portions of the glands may be implanted in other parts of the body under which condition the implant functions much as does the gland in its normal locus. Obviously, its function as a transplant will not be one hundred percent. as effective, in reference to its control of other glands, and functions, as would have been the control if the gland were located in the head with its normal portal circulation through the diencephalon.

The pituitary consists of two parts, an anterior portion whose extracts are only slightly toxic, while the extracts of the posterior lobe are highly toxic. The anterior lobe seems to be concerned with the general growth of the animal, and contains some fractional extracts which are known to activate the gonads. This extract acts synergistically with the internal secretion of either the testes or ovaries. Another fractional extract of the anterior pituitary stimulates and activates the thyroid gland, which gland as will be seen later, seems to be in general chemical control of the metabolism. Another known extract of the anterior pituitary controls the rate of fat metabolism in the body. If this extractive is not normally present fat tends to accumulate in certain parts of the body which will be discussed later as criteria for diagnosing under-activity of this portion of the division of the pituitary. Should this extractive be in excess in the secretion of the anterior pituitary, the individual will be excessively lean due to the excessive rate of metabolism and oxidation of fat within the system. Another extractive is known as prolactin which has to do with the secretion of milk by the mammary glands, the extractive acting synergistically with the mammary glands.

If the anterior lobe of the pituitary is over-active before the age of puberty there will be an excessive growth of bones, the genitalia, and of the voluntary muscular system, and if extreme may result in gigantism. If this excessive function develops after puberty it is obvious that growth of the long bones is no longer possible, and no increase in height would follow. In this latter instance the effect is more on the short, flat, peaked and irregular bones, more often those of the head and face, which results in the peculiar facial conformation found in acromegalia. Under-active function of the anterior pituitary before puberty results in a short stature, small bones, and an under-development of the voluntary muscular system.

The secretion of the anterior pituitary tends to decrease the ability of smooth muscle to contract and to *hold contraction*. Smooth muscles, it should be pointed out, are found in the blood vessel walls, the heart, intestinal tract, and in the interior of the eye. This lobe also tends to store bromides if administered medicinally thereby lessening synaptic response, but what seems to be worse, is that this storage directly affects the psychic responses of the patient, causing him to be apathetic, dull, and non-responsive to conditions met with in his environment. The anterior lobe exerts a rather powerful influence upon the ovulation in the female, and genesis of spermatozoa in the male, and to some extent controls ovulation and menstruation, thus determining the quantitative degree of fertility in the female. The secretion of this gland controls the libido in both sexes. Experimentally it has been found that direct stimulation by electrical currents of the anterior portion of the floor of the third ventricle in the brain seems to increase the secretion of this portion of the pituitary.

The posterior pituitary contains two well known fractions, one, pitressin, is a vasoconstrictor, and also controls the blood sugar level due to its effect upon carbohydrate metabolism. The other important factor is oxytocin, which causes the contraction of plain muscle, particularly those found in the uterus, blood vessels, gall bladder, and the intestines. A deficient functioning of the posterior lobe results in a marked obesity, which is usually of the very solid type, due to the retention of water and chlorides along with the deposition of fat. These depositions are characteristic in that they are around the pelvic and shoulder girdles, and on upper arms and thighs, the forearms and legs being relatively free of deposits. Incidentally, this type of fat accumulation is independent of the age of the patient, often being found in children.

Speaking of this gland as a whole, there are certain well

marked ocular symptoms which may result from hypertrophy of the gland, the most common one being hemianopsia which usually, in the early stages, affects color field vision, but later may affect form vision. If the enlargement extends much beyond the sella there is a possibility of interference with the function of the sixth cranial nerve resulting in a convergent strabismus. In a few instances the pressure may involve the third nerve resulting in a divergent strabismus. A still more extensive involvement may exert pressure upon the crura cerebri, causing faulty gait, with a positive Babinski sign. Epileptoid attacks are not uncommonly a result of such pressure. Other symptoms with which the patient may suffer are severe intractable headaches, usually affecting both templar areas, with an attendant vertigo. There may also be expulsive vomiting. The optic discs may become choked due to interference with circulation thereto accompanied by a progressive contraction of the visual fields, and ultimate optic atrophy. More specific ocular effects will be discussed later in this chapter.

The gonads seem to activate the general metabolism of the body perhaps mediated by the pituitary and its synergisitic glands.

In the male not much is known of the specific endocrine effects of the testes by direct experimentation. However, some things are known as a result of loss of, or destruction of the testes at different periods during the life of an individual. Early loss or destruction of interstitial function of the testes results in accumulation of fat around the hips, lower abdomen, between the shoulders, and in some instances results in an excessive growth of the long bones. The deposition of fat corresponds very closely in the male, so mutilated, to the normal distribution of fat in the female. Associated with these structural effects there is a failure of the voice to change from the treble to the normal baritone or bass of the adult male. For this reason castration has been practiced for ages upon boy sopranos in an effort to preserve the soprano voice for religious ceremonies in which the choral activities are all performed by men.

If the testes are removed or destroyed at or near puberty there is a failure in the normal development of the secondary sex characteristics, one of which was mentioned above, the voice change. Another notable change is the failure of the shoulder girdle properly to widen in comparison with the pelvic girdle. There is usually scanty if any facial hair, usually no axillary or suprapubic hair. The hirsute presence and distribution depends somewhat upon the age at the time of castration prior to the onset of puberty, which usually means that the mutilation to produce such end results must take place prior to eleven or twelve years of age.

If a loss of testicular interstitial secretion occurs after maturity, roughly from the age of fourteen years, about the only easily observed sign is a peculiar pasty cachexia, usually accompanied by diagonal wrinkles across the jaw line angling from above downward and forward. The effects under these conditions, are somewhat of a loss of stamina and the occurrence of early fatigue following physical effort. There is also a loss of the physical combative tendencies of the male.

In all three of the preceding age groups, having a loss of testicular function, there is a complete loss of libido, however, experiments tend to indicate that a proper administration of testicular substance, or an extract of testicular substance, may restore the libido, but, of course, would not produce active spermatozoa. Another effect of such administration would be an increase in the metabolic rate, a lessening of fatigue response, and, if continued long enough, an alteration of the deposition of fat in the body from that of the female type toward the normal male distribution and toward the normal hair distribution of the male.

It has been observed by *Harrower* that the thyroid gland in the castrated male never reaches the full developmental size found in the non-castrate male, perhaps accounting for the lower metabolic rate of the castrate.

In the case of the female gonads, the ovaries, considerably more is known of the effects of hyper- or hypo-secretion. The fractions of ovarian secretion have been more fully studied, the two principle fractions being folliculin and progestin. Folliculin affects the local development of the generative organs including the mammae. It must be noted, though, that folliculin in order to be potent requires an active anterior pituitary. In this respect they are definitely synergistic. Paradoxically as it may seem, folliculin also has an antagonistic effect upon the anterior pituitary despite the fact stated above that folliculin is inactive in the absence of the gonadotropic fraction of the anterior pituitary. Furthermore, failure of the anterior pituitary secretion disturbs the rate of ovulation, which is the specific reproductive function of the ovary.

Fridenberg[2] states that even the ancients recognized the significance of the gonads in the male and female as the determining factor in life, body form and character, and also the more obvious effects following castration to control sexual activity, and the operation of spaying to preclude pregnancy.

Failure of the ovarian hormones early in life results in a body structure in the female which closely approximates that in the male, characterized by excessive length of the long bones in the legs and arms in conjunction with a narrow pelvic girdle. The hands, wrists, ankles and feet are all thin and narrow with a tendency of the extremities to have a subnormal temperature, but they may remain dry due to a not excessive secretion of perspiration on their plantar surface. Another finding in this connection is a subnormal development of the glandular tissue in the mammae and a failure of the normal deposition of fat in the breast. Castration of the female, or loss of ovarian function once it has matured, results in an artificial onset of the menopause with the symptoms normally accompanying it, such as hot flashes, nervous irritability, and in some instances hyperthyroid activity, even to the point of a thyrotoxicosis. As in the castrate male there is a tendency to an easily produced fatigue with a prolonged recovery therefrom.

The adrenal gland consists of two structural parts, an inner medulla which embryologically develops from a like source as does the sympathetic division of the autonomic. The medulla of the adrenal activates the sympathetic and is activated by the sympathetic. The principle fraction of the secretion of the adrenal medulla is adrenine. In order for adrenine to be potent in the body it is necessary that the blood pH remain below 7.6. Should the blood pH rise above this level adrenine becomes inactive and incapable of producing its peculiar physiological response consisting of tonic vasoconstriction accompanied by a rise in blood pressure, a release of sugar in the blood, "goose flesh," tachycardia, blanching of the skin, dilation of the pupils,

[2] Fridenberg, "Endocrines, Vitamines and the Eye," Med. Rec., 148, 5, 173.

and in some instances, tremor. Adrenine has another property
which is made use of therapeutically in that it has the power to
relax spasms in the bronchioles such as are found in asthma.
Some authoities attribute the pale, boggy mucous membranes of
the nasal cavity to a failure of the constricting power of adrenine, due to a pH higher than 7.6. The symptoms described in
reference to the lungs and nasal cavities are those of the allergic
response by these tissues, which conditions are usually found in
individuals who have a lessened body acidity, technically called
an alkalosis.

Adrenine is constantly being secreted by the adrenal medulla
and this secretion is greatly increased during stimulation of the
sympathetic either directly, physiologically, or under emotional
stresses which involve the emotions of anger, fear, rage, and
which require some defensive or combative response in the animal.

Acids and calcium in the blood stream increase and activate
adrenine, whereas alkalies and potassium decrease adrenine activity and render it to be less potent.

During the period of increased secretion of the adrenal
medulla, due to sympathetic stimulation with the consequent
release of blood sugar, tachycardia, an increased respiration
rate, there co-exists a radical increase in the metabolic rate of
the individual. All metabolic processes are increased, and if
increased beyond the normal physiologic limits results in the
burning up of any stored food or fat in the body.

Since the adrenal medulla is sympathetically activated it is
possible to destroy the medulla or to destroy its functioning
power by section of the sympathetic fibers innervating it. Destruction of the adrenal medulla, strangely enough, is not detrimental to life, but it does render the animal, so mutilated, incapable of those reactions necessary for defense, and requires that
the animal be kept under environmental conditions in which
these reactions are not demanded. This is strikingly illustrated
in the carnivora who depend upon their combative ability to
secure food. Carnivora mutilated as above, are thereafter
unable by their own efforts to secure other animals for food.

A powerful adrenal medulla is thus seen to be a necessity for
strong skeletal muscle response. Also, for the recovery of

fatigued skeletal muscle to its normal functioning power. Here mention must be made of the fact that the extraocular muscles, those found in the orbit outside the eye globe, are strictly of a skeletal or somatic type, and their ability to maintain ocular fixation over long periods of time and to recover from the fatigue of such use becomes a function of the adrenal medulla acting in conjunction with a normal blood pH.

The adrenal medulla has as a synergistic gland the thyroid gland, and as antagonists the pancreas and the duodenum. A failure of the proper tonic antagonism between the pancreas and the adrenal medulla not uncommonly results in a hyperglycemia.

The outer shell of the adrenal gland, the cortex, activates the parasympathetic, and is activated thereby. The cortex is immediately necessary for life and any failure in its function alters the ability of the tissues in the body properly to utilize oxygen. The cortex also functions in maintaining a vital blood volume.

A fraction of the secretion of the adrenal cortex known as choline increases the activity of the parasympathetic producing a condition known as vagotonia and may decrease blood pressure either by decreasing the blood volume, but more particularly by a chemically produced vasodilation.

The cortex aids in the utilization of vitamin C and plays an important part in the metabolism of sulphur in the body. In young children an undue over-activity of the cortex produces precocious sexual development of the child accompanied by rapid growth of the genitalia with early evidence of libido. In this connection it mutually co-acts with the anterior pituitary.

The thyroid gland is not immediately necessary to life and releases several hormones, two of which are well known thyroxin and de-iodothyrin, the former contains a high percentage of iodine and is the active principle which mediates metabolism. The latter, also, contains a small percentage of iodine but is chemically and physiologically antagonistic to thyroxin. It is a notable observation that an absolutely vegetarian diet actually decreases the size of the thyroid and reduces its functional power.

The common test for thyroid efficiency is the basal metabolism test, formerly used as the sole diagnostic differentiating test for thyrotoxicosis. More recent research, however, lessens the im-

pact value of an altered basal metabolism rate, abbreviated
B.M.R., and it is now only considered diagnostic of altered
thyroid function when found in the presence of the other co-
existing clinical symptoms. The B.M.R. is usually made in the
fasting state by measuring the rate at which oxygen is utilized
by the body tissues. While this method is chemically accurate
it has been found that the B.M.R. is closely associated with other
clinical findings, notably the pulse pressure and the pulse rate.
Reed[3] has developed the following mathematical formula for the
calculation of the B.M.R. in the absence of the more elaborate
equipment which depends upon oxygen consumption. *Reed's*
formula is as follows: .683 times the sum of the pulse rate plus
.9 of the pulse pressure, to which is algebraically added the
quantity − 71.5. The final sum in this equation secured by the
use of clinical data obtained from the sphygmotonogram and
with a determination of patient's pulse rate, gives the B.M.R.
The findings are found to closely approximate those found by
the oxygen consumption method probably within a ± error of
two percent. Authorities are in relative agreement that the
physiologic range for the normal B.M.R. is between a + 10 and a
− 10. However, clinicians usually consider the range to be
safely physiological if it falls between a + 15 and a − 15. B.M.R.
determinations should not be undertaken during menstruation
or at the time the patient may be suffering from any of the acute
illnesses accompanied by fever.

A hyperthyroid condition in which there is an excessive secretion of thyroxin results in a high B.M.R. finding, accompanied
by an increased mental alertness with an "up-and-doing" socio-
logical attitude. Conversely, a low secretion of thyroxin as is
found in hypothyroid, results in the retention of water and
chlorides by the tissues causing an excessive body weight, dull
mental reactions, and sluggish response to stimuli, the state
being known as myxedema. Occasionally, there exists a congenital absence of the thyroid, resulting in a condition known
as cretinism, characterized by a lack of growth development,
idiocy and an inability for adaptation into a civilized environment. Some of these congenital cases may be helped by injec-
tions of thyroid extract or more simply by feeding the desiccated
thyroid glands.

[3] Reed, Goodale's ''Interpretation of Laboratory Findings.''

One of the accompanying effects of hypothyroid activity, probably due to the excessive retention of water and chlorides, is that the healing processes following trauma are delayed both as they apply to soft tissue and to bone repair and development.

The following effects have been experimentally determined as resultants of an increased thyroid secretion:

An increased utilization of oxygen
An increased blood sugar
Mobilization of water and chlorides
An increased pulse rate
An increased elimination of calcium and phosphorus
An increased tonus of both striped and smooth muscles
An increased tonic tension of the sympathetic, and
An increased nervous irritability with excessive reflex responses.

From the foregoing tabulation it can be seen that there may occur rather drastic ocular functional anomalies, particularly those depending upon a proper use of striped and smooth muscle, or proper sympathetic functioning.

The thyroid acts synergistically with the adrenals, the pituitary, the gonads, and the liver, and antagonizes the thymus, pancreas and parathyroid.

The thymus gland has not been studied as closely as have the other endocrine glands, presumably because this gland is peculiarly a gland of childhood and normally undergoes atrophy at or near puberty as the result of an increase of activity by the gonads and the thyroid. It is known, however, that early removal of the thymus causes an increase in the length of the long bones. Furthermore, the injection of thymus extract causes an increase in the growth of the external genitalia, hence the thymus undoubtedly contains a gonadotropic extractive.

Some authorities suggest a compensatory relationship between the adrenals and the thymus, because where a small thymus is found in an individual there are usually found large and massive adrenals and vice versa. This compensatory relationship seems important from the standpoint of this thesis because rapid and prolonged action of muscle tissue is dependent upon active adrenals.

The parathyroids have to do with, and control, the calcium

metabolism. A lack of calcium in the blood may result in tetany, which is a tonic spasm of somatic muscle. Lack of calcium may also cause irritation of the autonomic nervous system, and may cause trophic changes of the teeth, nails, loss of hair, and due to the deposition of calcium in the lens of the eye, may cause the senile type cataracts. An excessive secretion of the parathyroid produces reverse changes to those above mentioned, and because one of these changes is a lessening of calcium in the bones they tend to become soft and lose their form. The walls of the orbit of the eye are extremely thin bony plates which, should their calcium be removed by over-active parathyroids, would revert to thin cartilaginous plates which would then become elastic and might become distorted under increased muscle pull or tension as found in some of the phorias, resulting in their becoming actual tropias.

Bertram[4] insists that the pituitary gland governs the behavior of the animal as a result of his experiments performed on animals. *Fridenberg* in discussing behavior problems in relation to the endocrines finds that the whole complex of endocrine hormones, acting in conjunction with the autonomic nervous system, is markedly affected by the emotional and behavioristic responses of the individual. He continues, "These as we know, are elicited by sensory stimuli of which the most vital and probably the most numerous are the *visual stimuli.*" Continuing *Fridenberg* insists that the central gray brain centers, which have to do with facial expression as altered by the extrinsic ocular muscles, and those of the lids, must be further studied from a diagnostic standpoint. He significantly finds that the sensory end-organ, the eye itself, must be taken into account not only as to a source of secretory stimulation by the visual perceptions, but as a highly complicated organ eliciting nervous reflections in close inter-relatoinship with the autonomic nervous system, both its parasympathetic and the sympathetic components. He cites the *Aschner-Dagnini* oculocardiac reflex as one instance of the connection between the eye and the parasympathetic and comments upon the simultaneous use of adrenine, insulin or thyroxin in conjunction with efforts to evaluate this test.

4 Bertram, "Pituitary Behavior," S.N.L., Sept. 12, 1937.

Bieler[5] in his discussion of endocrine types describes a type of animal resulting from breeding experiments in efforts to augment adrenal strength. Such animals he names to be the wide-headed short horned bull, the wide-headed draft horse, and the English bull dog. He states that examination of the adrenal type human being indicates characteristics comparable to those found in the animals above named: to wit, wide heads, heavy massive necks, broad chests, large canine teeth, and teeth which are extremely hard and resistent to caries, coupled with a low forehead, heavy dark hair often curly, and a body with the "hairy ape" type of hair distribution. *Bieler* continuing his discussion of endocrine types makes the point that women belonging to the thyroid type, wherein the gland is over-active, have an increased rate in the menstrual cycle, sometimes as short as fourteen days, which is about half the normal cycle. He also states that the period of gestation of this type of woman is usually shortened, often as much as twenty days shorter than the normal two hundred eighty days. Babies delivered by such women are usually small and thin, very irritable, but otherwise apparently healthy. Mothers of the thyroid type usually have an excessive milk secretion and for that reason are sought as wet nurses.

In connection with pregnancy *Sobel*[6] makes use of an interesting observation in reference to the frenulum during pregnancy. He finds that under transillumination the pregnancy frenulum undergoes a change to a markedly pinkish hue, and that the large vessels, which are hardly visible in the normal frenulum, appear turgid having a pink capillary network between the vessles. As the pregnancy progresses the visible large vessels appear more numerous and the capillary network much more pronounced. He finds that these changes in part may also be observed during menstruation.

Draper[7] states that in hyper-pituitrism the teeth are large, widely separated, and that this is especially true of the middle incisors. He notes that overlapping of the middle incisors in individuals, accompanies, in the individual in which it is found,

[5] Bieler, "Endocrine Types," Clin. Med. and Surg., Apr. 1932.

[6] Sobel, "Changes in Tendencies in Uterine Conditions," Med. Rec., Mar. 1937.

[7] Draper, "Constitutional Iridiagnosis," Lea and Febiger, 1928.

a tendency to uterine malposition or fibroids in the uterus in the female, and to prostatic hypertrophy in the male.

In efforts to arrive at a diagnosis as to the present status of the endocrine glands the following observations should be made:

1. The distribution of fat, and whether the fat is hard or soft.
2. The quantity and distribution of facial hair in women.
3. The dermograph, which is a peculiar vasomotor response of the skin to friction.
4. Findings about the eye and bodily proportions.

Symptom findings about the eye embrace the following conditions which are attributable to the endocrines named in conjunction therewith:

1. The ability of the pupil to contract to light, with an inability to hold the contraction in the same light intensity for longer than approximately fifty seconds. These findings indicate a lack of proper adrenal function.
2. A sluggish rate of reaction of the pupil to light or convergence is also an adrenal anomaly.
3. Exophthalmus is an accompaniant of thyroid over-activity.
4. Ptosis may be interference with the cervical sympathetic, closely associated with thyroid and adrenal function.
5. Nystagmus may be a parathyroid involvement by reason of altered calcium metabolism with tetanic clonic contraction of extraocular muscles.
6. Convergent or divergent strabismus, as has been said before, may be due to pituitary involvement.

As mentioned heretofore, fat distribution on the body may be used in a diagnostic way for determining under-activity of certain of the endocrines. The fat distribution in the hypothyroid is distinctive in that there is an upper back pad of fat from about the seventh cervical to the fourth dorsal, sometimes called the "dowager's hump." The buttocks are well overlaid with fat which usually extends into the thigh, and has been given the name "fat breeches" or the "fat jodphur." The breasts are large and soft, but usually not pendulous. The legs and ankles are both fat, giving them a somewhat stumpy appearance. The facial expression is puffy and listless, the hair dry, while a lack of perspiration causes the skin to be dry and sometimes scaly. These individuals give the impression that the upper half of the body is much larger than the lower half. If the hypothyroid status occurs early in life it can sometimes be diagnosed by the very evident delay of dentitian.

In under-activity of the pituitary gland the fat is largely distributed around the mid-line and hips forming a fat girdle. The lower abdomen very often overhangs in an apron-like form extending below the symphysis pubis. The facial formation is round with a rather massive double chin. In hypopituitary males there is usually found a female breast-like fat distribution which may or may not be covered with hair. A significant criterion in the fat hypopituitary case is the fact that both the ankles and wrists are thin with no fat overlying them nor approaching them for a distance of roughly about six inches.

In the hypogonadal type of fat the distribution is in the upper one-third of the thighs with large fat pads over the trochanteric region. This latter distribution is facetiously called a "saddle bag" pad. These individuals have very slender fingers and wrists, thinking in terms of bony structure rather than fat distribution or lack of fat. Incidentally, the long bones in these cases are usually longer than the normal for the height of the individual.

Individuals in whom the adrenals are under-active show a full trunk as a whole with a full, puffy face, but the expression is alert as contrasted with the dull expression in the hypothyroid types. The upper arms are fat, as are the thighs, but the forearms and legs are usually free of fat, giving a somewhat "chicken drumstick" appearance to both. The wrists, ankles, hands, and feet are usually thick due to bone mass rather than a heavy deposition of fat. On the whole this type shows a stocky massive build on first impression and the hands and feet are usually warm and dry.

Impairment of the central gray function, since it affects all of the glands via the nervous system, causes a fat distribution similar to that found in the hypopituitary cases, with additional masses of fat over the buttocks, "fat breeches," such as are found in the hypothyroid types.

The eyes and associated tissues may be affected as a result of endocrine involvement and their consequent effect upon the autonomic nervous system. Enlarged glands in close association with blood vessels or nerve trunks may cause pressure effects which, if light, usually stimulate nervous structure, but if heavy pressure is exerted results in inhibition or even paralysis. A faulty endocrine system may also affect the eyes and their associated structures due to altered metabolism and the ability of the body to handle calcium and potassium particularly. The following table shows the effects of autonomic disturbances on the following ocular structures: the upper lid, conjunctiva, cornea, pupil, palpebral fissures, position of the eye globe in the orbit, and intraocular tension:

TABLE I
AUTONOMIC SIGNS

	Sympathetic	Parasympathetic
Upper lid	Retarded lagging	Ptosed-puffy
Conjunctiva	Vasomotor constriction	Vasomotor dilation
Cornea	Complicates conditions only	Dystrophic changes
Pupil	Dilated	Contracted
Palpebral fissure	Widened	Narrowed
Lacrymation	Increased	Decreased
Position of Eye globe	Exophthalmus	Enophthalmus
Tension	Hyper	Hypo

Increased activity of the sympathetic division of the autonomic may be caused by an overactive thyroid, an overactive adrenal medulla with a blood pH below 7.6, or an overactive posterior lobe of the pituitary.

Increased activity of the parasympathetic division of the autonomic may be caused by an over-activity of the parathyroid, with their increase of the metabolism of calcium, an increase of interstitial hormones by the gonads, or an over-active pancreas, which if the secretion of insulin is increased, radically lowers blood sugar.

It will be noted that the anterior pituitary and the adrenal cortex have not been listed in the next above two paragraphs. The evidence indicates that an over-active anterior pituitary stimulates the sympathetic, but this writer is unaware of any confirmed experimental evidence on this point. On the other hand, the adrenal cortex apparently stimulates the parasympathetic, while at the same time being activated by that division of the autonomic.

The following table, Table II, tabulates pressure symptoms by enlarged glandular structures and their effects, if intracranial pressure is produced, or if the pressure is in the cervical region. In this connection it should be recalled that intracranial pressure may cause, and usually does cause, impairment of the eye grounds as seen with the ophthalmoscope, or of the form and color fields as checked by the perimetric or campimetric methods.

TABLE II

PRESSURE SYMPTOMS

Where	Nerve affected	Degree of Pressure	Sympathetic	Parasympathetic
Intracranial	Oculomotor	Slight	Relaxant	Stimulant
		Great	Unopposed	Destroyed
Cervical	Sympathetic	Slight	Stimulant	Relaxant
		Great	Destroyed	Unopposed

Tabulated below, in Table III, are the ocular metabolic changes and functional changes which may accompany endocrine dysfunction:

TABLE III

METABOLIC EFFECTS ON EYES DUE TO ENDOCRINE DYSFUNCTION

A. ABNORMAL MUSCLE functions—erratic findings—
B. LIDS
    1. Lid lag—Von Graefe sign—
    2. Stare—Gifford sign—
    3. Slow winking—Stellwag sign—
    4. Lack of fixation convergence— }
       Moebius sign—Seemingly due to mechanical } THYROID
       pressure on eyeball in exophthalmus or ex- }
       cess tension of oblique muscles. }
    5. Also sympathetic action does the same.
    6. Paresis, ptosis, nystagmus, look for possible pituitary involvement.
C. CORNEAL AND CONJUNCTIVAL CHANGES
    1. Degenerative changes in hyperthyroid—due to exophthalmus.
    2. Keratoconus—adipose genital dystrophy, usually pituitary in origin.
    3. Keratitis—diabetes; Hypothyroid; phlyctenular type—hypoovarian. Conjunctival dryness.
D. CATARACT
    1. PARATHYROID PERINUCLEAR AND LAMELLAR, usually parathyroid, sometimes following thyroidectomy.
    2. DIABETIC—begins as fine, punctate, milky, white spots which soon coalesce to complete cloudiness—three to four percentum of all cases—
    3. SENILE—Hypothyroid—usually begins as cortical. Some say gonadal—controversial—
E. EYE GROUNDS
    1. Simple atrophy is pituitary in fifty percent. of cases.
    2. Choked disc is pituitary in only ten percent. of cases.
    3. Optic neuritis is pituitary in only twenty percent. of cases.
F. NERVE—usual changes
    1. Enlarged blind spot.

    2. Bi-temporal hemianopsia. Often Pituitary tumor.
    3. Unilateral hemianopsia—unusual.
G. FIELDS
    1. Pregnancy and menstruation may show *same* field changes due to
       pituitary enlargement.
H. TENSION
    Glaucoma
    1. Depending upon structure to cause glaucoma—contributing
       causes.
       a. Flat anterior chamber.
       b. Hyperopia.
       c. Large lens.
       d. Hypothyroid—erratic.
       e. Dished iris—convexly forward.
    2. Some Producing Causes
       a. Emotional stress.
       b. Fatigue.
       c. Cold.
       d. Thyroid—hyper.
       e. Climacteric women—old men—gonadal—
       f. Sympatheticatonia—extirpation of superior cervical gan-
         glion.
I. PIGMENTATION, ABNORMAL LOCATION OF, IN THE EYE OR
   ITS SURROUNDING TISSUES.

It is not intended that the foregoing discussion of endocrinology shall be a complete dissertation on this subject. Only those
facts which pertain to ocular problems or to general health problems which may be influenced by ocular stimulation have been
included. The significance of the practical application of these
facts will be considered later in Part III.

REFERENCES

Cushing, ''Pituitary and the Hypothalmus,'' Charles C Thomas, 1932.
Josephson, ''Glaucoma and Its Medical Treatment with Color,'' Chedney
   Press, 1937.
Hoskins, ''Tides of Life,'' Norton, 1933.
Badler, ''The Endocrines,'' Norton, 1933.
Harrower, ''Endocrine Diagnostic Charts,'' Harrower Lab. Publ., 1929.
Cole, ''Diagnostic Endocrine Syndromes,'' Cole Chem. Pub., 1931.
Timme, ''Biology of Individual,'' Williams and Wilkins, 1934, Ch. VII.
Lowenberg, ''Clinical Endocrinology,'' Davis, 1937.
Wolf, ''Endocrines in Modern Practice,'' Saunders, 1937.

CHAPTER X

BODY POTENTIAL, BRAIN WAVES AND ACTION CURRENTS

In the study of electro-chemistry it is well known that if dissimilar metals are immersed in a solution which attacks one of them but does not attack the other, or which attacks one of them more than the other, an electrical potential difference will be set up between the exposed ends of the immersed metals. The one attacked most by the chemical, called the electrolyte, possesses a negative polarity to the one less attacked by the electrolyte. Another not so well known phenomenon in electro-chemistry is that when two dissimilar salts or salt solutions are separated by a semipermeable membrane the set-up results in creating a potential difference between the two solutions. The electrical energy developed is a result of diffusion through the membrane and is known as the "diffusion potential."

A diffusion potential may also be set up between two solutions of the same salt separated by a semipermeable membrane provided that the concentration of one solution is greater than the other. The potential difference in the case of sols separated by a membrane depends to a large extent upon the chemical nature and the physical structure of the membrane, the latter in terms of thickness, density or compression, and other well known physical properties.

In the living cell in its aqueous environment there occurs a difference in potential between the inside of the cell and its environment. Structurally, the cell, as has been shown heretofore, consists of a living mass of protoplasm bounded by a semipermeable membrane, which separates its protoplasm from its food supply in solution in water. Strictly speaking, the surface of the cell is not a membrane, it merely being a difference in density of the cell protoplasm which bounds the cell and by some cytologists has been designated by the term "gel." In the case of the living cell the potential difference between its interior and its environment may be the resultant of two factors: 1. A difference in the solution on the two sides of the gel, and, 2. Certain increased activities within the cell which take the form of chemical action necessary for its metabolism, which chemical action causes the interior of the cell to become electrically negative to its external environment. The nucleus of the cell is positive electrically to the cell protoplasm.

Since the biological sciences deal with living things, no further consideration will be given to potential differences set up by differences in solutions in the two sides of the membrane, and the discussion will be limited to the biotic phenomena of the generation of electrical potential.

By the use of micro-electrodes it is possible to enter the protoplasm of a cell without greatly damaging its semipermeable boundary. When utilizing this technic it is found that a difference of electrical potential varying from ten to eighty millivolts is generated by the living processes attendant upon metabolism within the cell. By irritating the cell electrically, chemically, mechanically, or by the use of certain stimulating drugs, it is found that the potential difference between the interior of the cell and its environment is markedly increased. Certain inhibiting or depressing chemicals or drugs are found to so reduce metabolic processes that the potential difference slowly approaches zero. And in fact may reach zero if the chemical or drug used is poisonous to the cell protoplasm. In the latter instance when the potential difference reaches zero the death of the cell ensues. This is an irreversible process, except in instances wherein a proper potential difference is impressed within the cell, when compared with its environment, by using an external source of potential. In such instances, if cell protoplasm has not undergone too great disintegration, the death may be reversible and the cell continue to function, and if the electrodes are removed may complete its life cycle ending in cell division.

Not only do we find a difference of potential in the single cells, but we find differences in potential existing between the several tissues found in more complex organisms. In a given individual the potential varies between the several organs, but the greatest difference of potential exists between the nervous system and the organs themselves. *Telkes* and *Crile* found the greatest difference in potential to exist between the brain and the liver in man and the higher animals. *Telkes* says:

Life in the unicellar organism is an adaptively changing

difference in potential between the cytoplasm and the medium in which it exists and presumably, also, between the cytoplasm and its nucleus. In the lowest forms of multicellular organisms life is an adaptive difference in potential between the central nervous system and the rest of the organism; in the higher muticellular organism life is the adaptive difference in potential between the brain and the other organs and tissues.

Just as in the cell whose potential falls to zero the cell dies, so in man, when the potential difference between the brain and the liver falls to zero, organic and systemic function ceases. It is true, however, that the cessation of organic and systemic function does not immediately result in the cessation of cellular function, which latter usually continues until such time as the accumulation of the end-products of metabolism of the cells accumulate in their immediate vicinity in such concentration as to kill them. The time required for this effect varies, of course, with the individual and his state of toxicity at the time of failure of organic function.

An experiment illustrative of this death has been conducted by the writer using rabbits. The rabbit, under light anesthesia, is attached to the vivisection board, scalp shaved and incised and a skull trephine made. ·A platinum electrode, which has been previously tested for polarity difference with the companion electrode, is inserted into the brain. Following fixation of this electrode *in situ,* another is inserted into the liver and conductors are attached to the two electrodes and connected to a sensitive galvanometer or to a vacuum tube voltmeter. The meter will register a difference of potential as soon as the circuit is closed through a switch or key. Due probably to hemorrhage from the liver the vital forces of the rabbit slowly begin to fail. During this failure period, periodic closing of the key shows a gradual reduction of potential difference between the brain and the liver. Under anesthesia, as the experiment was conducted, it is found that the brain possesses a positive polarity to the liver's negative polarity. If the set-up is continued for a period of time there follows a gradual drop in potential, a point soon being reached where the meter registers a practical zero, shortly after which respiration ceases and the heart stops beating and to all intents and purposes the rabbit is dead.   If in making the set-up an external source of potential is shunted around the meter and the terminal from the brain before it enters the contact key, and if in this shunt circuit an open key is placed, it has been found that after the cessation of respiration with a stilled heart, depression of the key in the shunt circuit would impress a positive potential on the brain and a negative potential on the liver, and will cause a rather sudden beginning of respiration which is later followed by regular pulsations of the heart, thus restoring the rabbit to a living status, which as a rule, of course, is short, due to liver hemorrhage above mentioned.

Telkes[1] reports that increasing the ether anesthesia during similar experiments to the above produces a steady drop in the potential difference and eventually results in death of the animal.   We have, therefore, two possible causes of death in the above experiments, the one mentioned, hemorrhage from the liver, and the other produced by the anesthetic itself.

Burge, Wickshire and *Shamp*[2] found that while the brain potential was positive to the liver under anesthesia, as consciousness returned, the potential reversed until the cortex became electro-negative.   It is, therefore, reasonable to assume that in the waking state a patient under examination would exhibit polarity such as that found after the evanescence of general anesthesia, i.e., brain negative, liver positive.   *Telkes*[3] in probing the brain using her special technic, finds that there are differences of potential in the form of standing charges between different parts of the brain itself, regardless of the difference in potential between the brain as a whole and the rest of the body.

Rosenthal quotes *Bangs*, who in turn is quoted by *Livingston*,[4] to the effect that the normal current in the human body in health approaches .08 to .1 µ amperes with a potential difference ranging from four to five millivolts.   Why this potential difference is so much lower than that of an individual cell is unknown to the writer unless perhaps it may be due to the parallel contact connections of the multiplicity of cells forming such a complex body as that found in the primates.

[1] Telkes, ''Phenomena of Life,'' Norton, 1936, p. 272.

[2] Burge, Wickshire, and Shamp, Anes. and Analg., 15, 261–267.

[3] Telkes, ''Phenomena of Life,'' p. 264, Norton, 1936.

[4] Livingston, ''Reciprocal Link in Life,'' Pub. Author, 1933, p. 86 *et seq.*

During the process leading to death the brain loses conductivity and the liver gains in conductivity. In fact it has been found that the potential reverses polarity shortly after death followed by a gradually increasing reverse potential difference for a period of time approximating twenty minutes, which is followed by a rapidly lessening of potential to zero. In the writer's experience when this latter potential difference reaches zero there is no possibility of restoring organic function by the re-impression of a positive potential on the brain, and a negative potential on the liver by using an external source, as detailed heretofore.

It has also been found that a potential difference exists in plants just as in animals, the tops of plants or trees carrying a positive potential to the roots which are negative. An interesting phenomenon in this connection is that with the electrodes in place for determining the polarity, the impression of an external *reversed* potential soon results in the death of a plant or tree. Obviously, the potential impressed must have a quantitatively higher value than the normally existing potential difference in the plant.

Since the maintenance of a potential difference is dependent upon insulation, it would appear that insulation of the animal body from the earth might be conducive to the maintenance of its functions and its life. *Vles,*[5] of Strausburg, reports on his experiment in which new born infants were kept in insulated cribs. Experiments were conducted with identical twins, one being placed in an insulated crib and the other in the regular uninsulated household crib. *Vles* reports that the infant in the insulated crib grew and developed much more rapidly than the uninsulated infant. Thus, by preventing leakage of body potential to earth the metabolic processes of the infant in the crib were so increased as to show a measurable quantitative difference between it and its identical twin.

The conclusion seems obvious that to maintain life and adequate function it is necessary to see that an optimum potential difference is maintained between the brain and the other organs and tissues. *Marinesco*[6] presents an interesting diagram in his

[5] Vles, Modern Mechanics, 5, 19.
[6] Marinesco, Sc. Mech., 4, 34.

article in which he shows a diagram of an ordinary electrical battery in its external circuit, and in a parallel diagram shows the human body. In the latter diagram the facts heretofore presented are schematically clarified.

As mentioned above when cells are stimulated the potential difference increases. A stimulant is defined by *Thompson* of McGill University, as follows: "Any change in environment which produces an active reaction in a cell is called a stimulus." Attention is directed to the phrase "active reaction" in the foregoing definition. Any application made to a living unit which results in a response as a result thereof, becomes a stimulus. Conversely, if there be no response to the change in environment the change cannot be classed as a stimulus.

This reaction of the cell to stimulation has been called the "action current" or "the current of stimulation," and is one of the attributes of all living things. The ability to continue to live is merely a constant series of reactions to multiple and varying stimuli. Life, therefore, might be defined as a reaction to environmental conditions or more properly, the power to react to varying environmental conditions.

Burns and *Lam*[7] of Yale have demonstrated that alteration of physiologic activity causes characteristic changes in the electrical potential, both quantitatively and qualitatively under the stress of activity. *Koltzov*[8] is reported in Scientific Progress as having proved that spermatozoa, which are known to contain an odd or even number of chromozones, can be separated by an impressed potential in the solution in which they live. His finding is that those spermatozoa which will produce females carry negative charges and are attracted to the positive pole, *contra* those spermatozoa which carry positive charges and produce males are attracted to the negative pole. A practical application, in Russia, of *Koltzov's* findings was the separation of spermatozoa for artificial insemination of ewes so that the number of ewe lambs could be increased above the normal probability.

Max[9] has demonstrated that emotion, worry and fear produce active action currents in the nervous system of comparable de-

[7] Burns and Lam, Science Digest, Mar. 1937.

[8] Koltzov, Scientific Progress, Sept. 1934.

[9] Max, S.N.L., Sept. 16, 1933.

gree to those produced by sensory stimuli. One of *Max's* experiments was conducted with deaf mutes whose habitual form of expression is by hand and arm movement. He found that these deaf mutes showed quite marked action currents in the hands and forearms, during dreams, the existence of which was also demonstrated by the Berger rhythm which will be discussed later.

Loomis[10] demonstrated action currents in *facial* muscles, jaw muscles, and the muscles involved in swallowing in the throat, in association with emotional stresses, which action potentials were present separately and were distinctly different from the brain waves and were in no wise synchronized therewith.

Smith[11] and *Ruckmick*[12] have for some time been measuring potential differences resulting in individuals under emotional stresses. Both of these men find, when proper electrical circuits are used, that very definite changes in reading of the galvanometers take place during emotional reactions. *Smith* uses zinc electrodes covered with wash leather soaked in strong salt solution, and places one electrode on the palmar surface, and the other on the back of the hand with circuit through suitable potentiometers and meters. With this arrangement he finds a definite lessening of resistance through the hand when the patient reacts emotionally to certain words in a word association test or to other situations which arouse his emotions. *Ruckmick's* findings are in agreement with these and are probably more accurate quantitatively, due to differences in the circuits used by these two investigators. The above are demonstrations of the psycho-galvanic reaction.

Whether the above potential difference as recorded is a result of action potentials in the same direction as the impressed current, thus giving an additive result on the meter reading, or if the change in meter reading is due to lessened skin resistance during emotional stresses, is still a moot question. The evidence seems, however, to indicate the latter contingency as the true explanation of the phenomenon because of increased secretion in the palms of the hands, soles of the feet, and in the axillary region

[10] Loomis, Science, 81, 2111, 579.
[11] Smith, ''Measurement of Emotion,'' Paul, Trench, Trubner, 1922, 170.
[12] Ruckmick, ''Psychology of Feeling and Emotion,'' McGraw-Hill, 1936.

during stimulation of the sympathetic division of the autonomic. Since emotions force this division into action it appears that the evidence in favor of lessened resistance should be given more weight than evidence so far adduced in support of the action current theory in the same direction as the impressed potential.

Miles shows that the human eye as a whole shows electrical potentials just as do other tissues. The anterior portion of the eye carries a positive charge and the back of the eye a negative charge. These potentials can be lead from nearby tissues by the mere act of attaching electrodes near the anterior and posterior poles of the eye, which electrodes in turn are connected to a vacuum tube voltmeter. He found that the mere turning of the eye in the orbit causes a potential difference which remains higher and constant so long as the eye remains fixed in its position. The potential difference for a thirty degree deviation with fixation ranges from .2 to 2 millivolts for each eye. Furthermore, another interesting fact is that the potential difference may not be the same for each eye. It makes no difference during the execution of the experiment above whether an equal quantity of light falls into the eye in the original fixation position as compared with the secondary position. If the light differs in the two positions there is a slightly perceptible difference in potential due to this difference in light. Refractive defects apparently have no effect upon the change noted.

In the living body a specialized group of cells which are highly irritable to stimuli have to perform the function of correlating the multiplicity of responses into a harmonious whole. These specialized cells and their connecting protoplasmic links constitute the nervous system which has been heretofore discussed. It should be recalled here that the unit of this nervous system is the neuromere, consisting of a receptor, an afferent conducting path, a nerve cell and its synapses, an efferent conducting path, and finally an effector, which may be muscle tissue, gland tissue, or even other nerve cells. Stimulation of a nerve cell in this neuromere dispatches a nervous impulse over the connecting link to the effector or mediator. It is not necessary, however, for the nerve cell to discharge over the efferent path because it has been found experimentally that a stimulus applied directly to the efferent path will cause a response of the effector to which it goes. Such a stimulation of the path sets up an action current in the fibers which travels both ways from the point of the application of the stimulus.

The chemicophysical change responsible for the transmission of a nerve impulse takes the form of an increased utilization of oxygen by the fiber and the stimulus itself travels along the fiber as a wave of depolarization of the difference in potential normally existing between the center of the fiber and its surface, the center of the fiber normally having a negative potential to its external surface. Any stimulus which will permit a leakage of negative ions through the semipermeable boundary of the nerve fibers will result in depolarization, by reason of the neutralization of the normal positive surface charge by these leaking negative ions. This wave of depolarization travels along the nerve fibers at a velocity somewhere between twenty-eight and one hundred meters per second in the fibers constituting the cerebrospinal system. In the autonomic nervous system the velocity of the depolarization wave varies between .3 of a meter to approximately 5 meters per second. The obvious conclusion from this fact, supported by experimental evidence, is that the cerebrospinal system is set up for immediate or quick action in response to environmental demands, while the autonomic, which governs the more sluggish vital process, is not so quickly responsive.

A nervous impulse is not a continuous flow of energy following the application of a stimulus, but is an intermittent discharge of energy, roughly comparable to the quantum flow of energy in radiant energy. *Swindle*[13] has resolved virtually all of the biological phenomena to the quantum theory basis as a result of some investigation at Marquette University Medical School. He applies the quantum theory to the response of nerves, muscles, glands, and even to emotional responses. *Adrian*[14] hypothesized that the nerve impulses in all probability consist in a segregation of ions in parts of the fiber, which accumulation upsets the balance of processes naturally present in the resting state. He cites that a stimulus, to cause this segregation or accumulation, requires that it have a rather sudden application because a

[13] Swindle, ''Quantum Reactions and Associations,'' Badge, 1922.
[14] Adrian, ''Basis of Sensation,'' Norton, 1928, 19 *et seq.*

gradual application of electrical current will not produce an impulse of the nerve. To state his proposition in other words would be to say that the effectivity of a stimulus depends upon its rate of change as well as the intensity of its application.

If the rate at which the stimulus is applied and its intensity be sufficient to start an impulse he finds that the impulse is in the form of a volley of unitary quantities of energy, and that the frequency of the volley is somewhat proportional to the intensity of the stimulus, and in part to the suddenness with which it is applied. The responses cited by *Adrian* were those of the usual diphasic nerve responses which are secured when two electrodes are in contact with the surface of the nerve fiber. This form of response results when one electrode is on a normally polarized portion of the nerve and the other electrode sufficiently removed from it proximally to the stimulus, so that the wave of depolarization impinges upon one electrode with a time intervening before the wave reaches the second electrode. By this means he found that the wave of depolarization measured on the surface of the nerve is definitely negative to the normal resting state to the surface of the nerve which is positive. The result of this condition is that the current of action passes through the galvanometer in one direction when the wave of depolarization reaches the first electrode, but that it reverses as it passes through the galvanometer in the opposite direction when the wave reaches the second electrode. This change of direction constitutes the diphasic response.

On the other hand, if one electrode is placed into the end of the nerve fiber, and the other on its surface at a short distance from the cut end, the response is monophasic, in that the current only passes in one direction through the galvanometer at the instant the depolarization wave reaches the electrode on the surface of the fiber.

Following the passage of the wave of depolarization down the nerve fiber there is a period of about two to three thousandths of a second during which it is impossible to stimulate the nerve. This is known as the absolute refractory period of the nerve. If a time lapse of two hundredths of a second follows the passage of a wave of depolarization the nerve again returns to its normal state of excitability up to about ninety-five percentum of its normal. Addition of these two time elements shows that from twenty-two to twenty-three thousandths of a second is the shortest interval between two stimuli which may be applied to a nerve fiber and get an approximation of normal response, up to about ninety-five percentum of the normal. Incidentally, it is the sum of these refractory periods which make the ordinary motion picture possible. If two pictures are thrown on the screen with time intervals of less than twenty-three thousandths of a second they appear as one picture, but since the relative positions of objects in the several pictures are different in each picture, they thus create the illusion of motion in the observer.

Another physiological finding is that the intensity of the response of the effector in any efferent path is not the result of the intensity of the variation of intensity of the nervous impulse, but is a function of the *frequency of the volley* of the impulse. As can be seen from the foregoing discussion of the refractory periods of the nerve, this volley might theoretically have a frequency, if consideration only be given to the absolutely refractory period, of from three hundred and thirty-three to five hundred per second. Insofar as this writer is aware no nerve frequency measurements have ever been obtained above three hundred per second. *Fry*[15] of Ohio State University, reported to the writer that he at one time secured a frequency as high as one hundred and fifty per second using a rabbit's eye as a receptor and the optic nerve as the other pole in the experiment. Due to the synaptic delay in the bipolar cells and ganglia in the retina it appears that *Fry's* finding is in close agreement with a possible theoretical maximum obtainable.

Although the receptor begins its action instantly upon the receipt of the stimulus, it should be noted that there is a very appreciable lag in time between the incidence of the stimulus and appearance of the impulse in the nerve fiber. This lag approximates one-twentieth of a second for the ordinary sensory receptors. In the eye, however, this latent period may be considerably less. *Bishop*[16] cites that there is an appreciable delay between the incidence of the stimulus upon the retina and an

[15] Fry, Research communication.
[16] Bishop, ''Further Study of Elements of Optic Pathway,'' Am. Jr. Oph., 11/34.

appearance of electrical disturbances in the cerebral cortex. This delay he opines is caused by the lag at synapses in the bipolar cells, in the retinal ganglion cell layer, at the thalamus, and even in the cortex itself. Regardless of this delay the frequency of the impulse still varies with the intensity of the stimulus and, as has been pointed out above, the functional response is determined by the frequency of the impulse. The writer's experience dictates the conclusion that not only does the intensity of the light stimulus govern the frequency of the response but the frequency of the light also has an effect upon the frequency of the nervous impulses from the eye. The reasoning from experimental evidence is that the higher frequencies are more capable of ionization of the photochemical substances in the retina than are the lower frequencies. The result would be a greater potential development on the retina under higher frequency due to this ionization than it would be possible to obtain under the incidence of lower frequency light. And since it is this development of potential in the retina which is the immediate cause of the discharge by the optic nerve it follows that variation of light frequency directly varies nerve impulse frequencies.

In reference to the energy required for the emission of low frequency red light and high frequency violet light, an application of the *Einstein* law, $U = NhF$, where U is a constant representing the energy evolved during the chemiluminescent reaction of one gram molecule of the solution ; N, the number of molecules ; h is Planck's constant; and F the frequency of the light. By the application of this formula *Harvey* calculates that the reaction to just produce a visible red will require thirty-seven thousand calories per gram molecule, and seventy-one thousand calories per gram molecule to produce violet light. It should be noted that the requirement is almost double for the emission of violet light as contrasted with the requirement of just visible red light. These figures correspond very closely to the octave law.

It is a well known psychologic fact that if the eye steadily fixates a source of light that the apparent intensity of the source slowly decreases. This result is attributed to the bleaching of the photochemical substances in the photoreceptors in the retina. It is interesting to note in this connection that if the eye does not maintain *absolute* fixation on a source of light, so that the image of the source falls momentarily upon a point on the retina, which prior thereto only received scattered light outside the focal pencil, that this eye shift causes a perceptible sense impression of increase in intensity. For this reason, in developing a technic for the practical application of various frequencies of light the writer used a virtual light source much too large to permit constant fixation upon it as a point. The result was that as the visual axes move about over the apparent source, new points on the retina will fall under the focal pencil, producing a constantly apparent varying of the intensity of the stimulus. Since eye movements are quite rapid, the changes of apparent intensity are sufficient to excite additional impulses after the manner of the application of a virtually instantaneous stimulus. It must here be pointed out that the character of the nervous impulses over the optic nerve from the retina are not different from the character of nerve impulses from other sensory receptors. This being true, the interpretation of impulses from the eye must lie within the brain itself and not in any altered or differing kind of nerve impulses, other than perhaps their frequency, caused by the incidence of light or the *kind* of light upon the retina.

Owing to the lessening of the rate of discharge from a constant stimulus point on the retina anywhere within the visual field, we find that any movement of the eye which carries the visual axis over a source of light will give a greater excitation pattern of response to light, shade, or frequency, if the source is stationary and the eye is left free to shift its visual axis over the source. Such a result will not take place if the source is in motion and the eye constantly fixates the *same* point in the moving series.

Crile in summarizing results of investigations in his laboratory says in respect to sensory receptors that the special senses which are affected by light waves, sound waves and chemical influences are the primary controllers of the adaptive variations in the rate of body oxidation, which is just another way of saying that control of the wave-length of the absorbed emission controls protoplasmic tissue during oxidation. The chief organs involved in systemic oxidation are the brain, the thyroid, and the adreno-sympathetic system. From the foregoing it can be seen that the stimulus of light upon the retina may directly control or alter general metabolic processes. It should be noted in this connection that excessive stimulation of the receptors of the special senses, even common sensations, when stimulated singly or in any combination, causes excitation, followed by depression, and death if the stimulus is too great or is continued too long.

In order to study this possible mechanism an analog may be set up involving the photo-electric cell circuit as a parallel to the human body. The following figure VI is a circuit diagram showing the essentials for a photo-electric cell set-up.

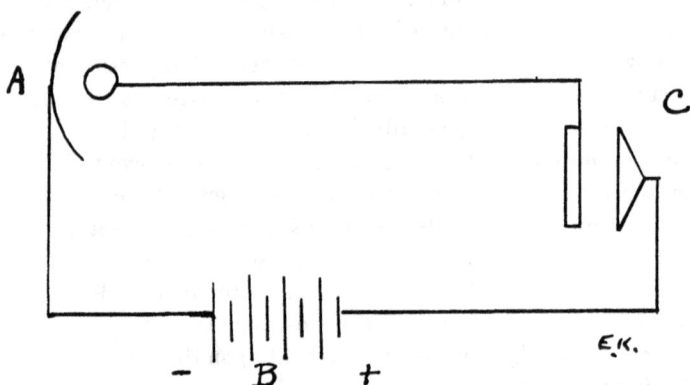

Figure VI.

"A" is a photo-cell with its emission surface connected to a negative potential in series at "B." "C" is a Zeleny electroscope in series with the positive lead to the collector of electrons carrying a positive charge within the photo-cell. With this set-up, and an alkaline metal emitter in the photo-cell, it is found that when the set-up is in a place of semi-darkness, just enough to see and count the motions of the detecting electroscope foil, that there will be a certain number of oscillations of this foil due to a very low grade quantitative emission of electrons from the surface of the alkaline metal. If red light is now thrown through the cell onto the alkaline metal, and the oscillations of the electroscope counted, it will be found that they number virtually the same as in the semi-darkness prior to exposure to red light. If now a mid-spectrum color is thrown upon the alkaline metal a perceptible increase in the rate of oscillation takes place. If the extreme opposite end of the spectrum, in the violet range, is now cast upon the alkaline metal in the photo-cell, a marked

increase of oscillation is seen, which approximately quadruples
the rate of oscillation found in semi-darkness or under red light.
Under the conditions of the experiment, since at each contact of
the oscillator with its charging plate there will be a drain on
the battery, this drain will be greater when under the influence
of violet light than for red light or semi-darkness. The time for
"running down" the battery is therefore much less under the
increased ionization produced by the high frequency light, until
eventually the battery would show zero potential in a much shorter
time than it would under semi-darkness or under red light.

Figure VII is a diagram showing the body with an eye, a brain,
a liver, with the choroid containing its highly vascular bed, and
the highly vascular negative liver, and the internal circuit of the
body through the electrolyte, in this case the blood stream. It
will be noted that here we have an electric circuit comparable to
Fig. VI with similar phenomena as to "battery run-down."

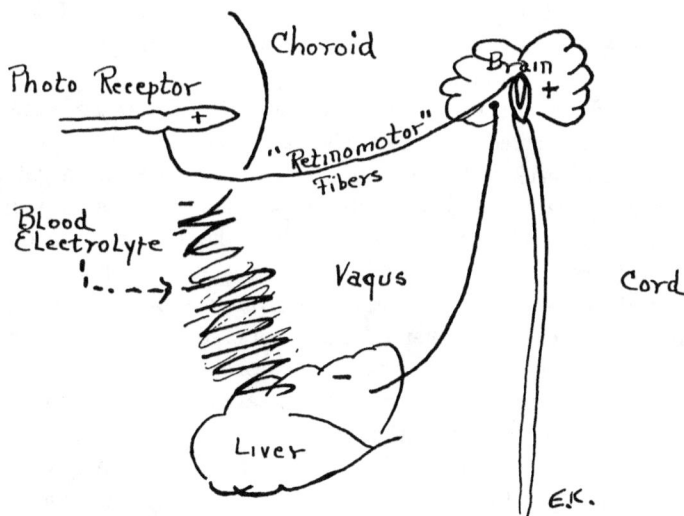

Figure VII.

Fisher implies that there is a standing positive charge in the
retina in the following observation: "An initial negativity immediately follows exposure to light." It is not unreasonable to
believe that a standing positive charge exists on the retina due
to the fibers running from the brain to the retina which have

been mentioned heretofore, and which have been called "retino-motor" fibers by various authors. If this standing charge on the retina, with its rods and cones, is placed in the path of a light, and if the choroid electrolytically carries a negative charge, we have a condition analogous to that shown in Figure VI. When light of low frequency is thrown into the eye the rate of ioniza-tion, choroid to retina, would be low, and the rate of synaptic discharge to the brain would also have a low frequency, com-parable to the low frequency oscillation of the electroscope, Figure VI. When high frequency light passes through the retina to the choroid, we would have a higher degree of ionization, with an increased build-up of negative charge in the retina, with an increase in synaptic discharge to the bipolar cells and the gangli-onic layer of the retina enroute to the brain. It follows, there-fore, that the potential difference existing between the brain and the other parts of the body should tend to fall under high fre-quency light at a greater rate than under low frequency light. Obviously, a similar phenomenon might be present under high intensity light of mixed frequency.

In an effort to confirm this hypothesis the following experi-ment was performed: An electrode was placed in the brain of a rabbit and another in its liver, in series with the detecting galva-nometer of the vacuum voltmeter type. After the set-up had become stable, red light was thrown into the eye of the rabbit, while all other light was excluded, and a notable increase of potential between the brain and the liver was observed on the meter indicating that either the stimulus of the light had raised the potential directly, or what is more probable in view of the above, the red light had decreased ionization and potential loss between the choroid and the retina, thus enabling the vital processes of the animal to accumulate a higher potential differ-ence. Following this observation high frequency light, at the other end of the visible spectrum, was thrown into the eye and a decrease in potential difference shortly began to register on the meter. This latter phenomenon was interpreted as indicating that due to the increased ionization, choroid to retina, the leak in potential between the brain and the liver had been increased, with a depletion of the potential difference at a greater rate than it was possible for the vital processes of the animal to replenish or to maintain it. In regard to the preceding experiment, performed a number of years ago, prior to the writer's knowledge of approaches by others to this problem, the writer feels gratified to note that *Crile's*[17] findings *supra,* confirms the present writer's, even though *Crile's* approach was by an entirely different route than that used by this writer.

It is a well known fact that photochemical effects are reversible in darkness. This phenomenon is found in the human eye as a re-development of rhodopsin in the rods either as a replacement process or by reversion to the positive potential on the retina as implied by *Bishop.*

Histologically, the retina consists of two types of photoreceptors, the cones and the rods. To the cones are attributed the function of form perception under high degrees of illumination while to the rods is assigned motion perception, and to seeing in very low degrees of illumination. A cross section of the optic nerve shows it to contain fibers of differing sizes, and it is found that the larger fibers, when stimulated, cause greater alterations of brain currents than when the smaller fibers are stimulated at their outer ends. A reasonable assumption might be that the larger fibers are rod fibers and the smaller fibers, cone fibers. Phenomena cited by *Houston,*[18] those of lateral conduction across the synaptic layers in the retina, seem to support the foregoing assumption in that a multiplicity of rods, due to crossed or lateral conduction across the synaptic layers, might result in the excitation of a single large nerve fiber. The picture is that of a multiplicity of rods having their impulses by lateral transmission across the retina, delivered to a single large fiber. *Eldridge-Green* is variously quoted as having stated that the rods are not organs of vision as such. Their function from the foregoing appears to be that of supporters of the cones in low degrees of illumination as a result of the summation effects of multiple rod impulses. This probably accounts for the fact that as we proceed from the macular region of the eye toward the periphery, with a gradual increase of the number of rods over the number of cones, it is found that perception of color decreases and finally disappears. *Ricco's* law of foveal vision when applied to a given

[17] Crile, ''Phenomena of Life,'' Norton, 1936, p. 362.
[18] Houston, ''Vision and Colour Vision,'' Longmore, 1932.

quantity of light with its focus at a point, and then thrown out of focus until it covers a retinal area of four million times that of the original area, produces a practical constant in the nervous impulses thereby elicited. The only way this phenomenon can be accounted for is on the basis of *Houston's* report of lateral conduction in the reticular layers causing the whole quantity of light absorbed to produce response as a unit quantity regardless of the area.

From the foregoing it appears to be clear that the choroid has some function to play in the act of seeing, if only to establish an electrical means for the execution of the functions of the photo-receptors in the retina under the influence of light. A further fact leading to this conculsion is that if the retina be separated from the choroid it becomes inoperative and incapable of excit-ing impulses from the detached area. Such a detached bit of retina very soon bleaches to a pale semblance of its former color and is practically devoid of rhodopsin.

Like most organic dyes, rhodopsin bleaches when reduced by the action of hydrogen and has its color restored upon re-oxida-tion. Electrochemically, hydrogen carries a positive charge in its ionized state, consequently it would be attracted to a negative polarity. As we have seen above the retina changes from posi-tive to negative upon light stimulus. If this negative charge attracts hydrogen, rhodopsin is bleached and a portion of the retina becomes less sensitive to light. It seems theoretically possible that there may be an optimum degree of this bleaching beyond which the potential might reverse under a continued absorption of light. It is known that there is such a potential reversal, and that it is accompanied by a considerably lessened frequency of nervous impulses passing over the optic nerve into the brain. *Contra,* oxygen carries an electrochemical negative charge and would be attracted to a positive charge on the retina of the standing charge type. If the retina is kept in darkness, rhodopsin has its color restored, and in view of the foregoing, it is not inconceivable that the ionized oxygen attracted by the positive charge is responsible for this reversed condition. It has, also, been shown by a number of investigators that Vitamin A in proper quantities is a necessary constituent of the blood stream to enable the regeneration of rhodopsin to take place.

Furthermore along this same line of thought, investigation shows that the scotoptic retina has an alkaline chemical reaction, whereas the photoptic retina has an acid reaction. The measure of acidity chemically is determined by the mobile hydrogen ions in solutions. This latter fact being merely a confirmative fact to the foregoing contention.

A number of years ago it was first shown by investigators at Kaiser Wilhelm Institute, notably by *Krummueller,* that the nerve currents in the brain are essentially electrical. Since this finding a rather large number of investigators have been delving into this field. *Jasper* and *Carmichael*[19] of Brown University, reported that they had succeeded in recording these fluctuating brain currents by using a recording oscillograph. In some of *Berger's*[20] work he isolated a wave which is relatively constant in all sleeping brains. The frequency of this wave which he called the α rhythm, approximates nine or ten per second. A number of conditions are known to alter this rhythm, and in order to establish a criteron for the evaluation of this rhythm in any individual, *Berger* established an index factor. His index is the sum of the α waves that are recorded in thirty seconds of time.

It has been shown that there are only a few α waves in the patient's index in the presence of deep psychological drives, such as powerful emotions. On the other hand, the α index is considerably higher and has greater amplitude, i.e., higher voltage variations, when the individual is psychologically passive or asleep. These α waves are apparently in nowise associated with intelligence, but, *contra,* are a measure of the psycholgic instability or the presence of emotional drives. *Berger* also isolated another commonly present rhythm which he called the β wave, which has a frequency varying from twenty to twenty-four per second. He early discovered that the brain waves which were present during sleep and with the eyes closed, the α rhythm, disappeared or appeared as a modulation of an action current when the eyes were used to view some object, or *light was thrown into the eyes. Loomis* confirmed this finding at a later date and also found that sounds which were *not sufficient to awaken* a sleeper caused the cessation of the *Berger* α rhythm or caused it to appear

[19] Jasper and Carmichael, S.N.L., Jan. 19, 1935.
[20] Berger, Arch. of Psych., 1929.

as a modulation of a larger wave. These findings tend to indicate that the *Berger* rhythm, the α rhythm, may be associated with the rhythmic discharges of nervous energy which maintain vital functional activity. The occurrence of a sudden sound during sleep, or the flashing of a light in the face would obviously arouse the defensive mechanism in the body, which is under control of the sympathetic division of the autonomic, and the normal dynamic antagonism between these two divisions may probably be the cause of abolition of the α rhythm.

Another of the findings by *Loomis,* reported at a later date, is that if a subject in a dark room, with the eyes open has a *light flashed into the eyes* there will be set up in the wave trains, special trains not unlike those following the flashing of a light on the face of a sleeper, even if the light so used was *so faint* as to require thorough dark adaptation of the eyes before the experiment. The conclusion reached is that the intensity of the light is *not* a factor in ocular sensations as a result of light, the real factor being the present state of the receptor and the conduction paths.

Hoagland[21] reports a very definite correlation between the *Berger* rhythm and body temperature. His finding is that an increase of body temperature of 6 degrees F.—3.5 C.—increases the α rhythm approximately thirty-five percentum. Since an increase of body temperature of 6 degrees is due to an increase of metabolic function controlled by the thyroid and the sympathetico-adrenal system, it is obvious that this increase of rhythm is either caused by activity of this system or is normally associated therewith.

At Harvard Medical School it was shown that the amplitude of the α rhythm, i.e., its voltage, rises rapidly when the skin is chilled, and also during the rigor in the early stages of infectious conditions. Incidentally, that department also discovered that there was increase of voltage of brain waves, as much as three thousand percentum, during an epileptic seizure, during which all the skeletal muscles of the body are in clonic contraction. It seems, therefore, from these findings, and those of *Jasper* that for the production of regular amplitudes, i.e., voltage deviation, and

[21] Hoagland, Science, 83, 2143, 84.

regular rhythm, the required condition is complete relaxation and quiet for the subject during the experiment.

Lindsey[22] found a definite relationship between the early brain development and the α rhythm with the *inception of the sense of sight* in infants at about the age of four months, when the functional activity of the visual area in the occipital lobe of the cortex becomes *first active*. It should be noted here that visual impressions pass through the thalamus enroute to the cortex which may account for *Gibb's*[23] discovery of a correlation between the α waves and the act of respiration which is controlled in this region of the brain.

Girard[24] succeeded in checking upon the optic radiation in the brain by probing its surface, using an electrical circuit, and found that the *surface potential of the brain* changed in very definite locations of the brain when *light is thrown into the eye*. *Girard's* work thus becomes a confirmation of that of *Bishop* reported above. *Girard's* experiments, by making use of the delivery of impulses to the cortex, located the regions into which visual sense impressions are delivered, his findings being an electrical confirmation of the location of these centers as previously determined by the older methods of using degenerated fibers for tracing nerve paths through the brain. *Girard* found that these impulses from the optic nerve could be traced through the *chiasma* into the *thalamus,* into the *midbrain,* and into the *cerebral cortex.* He thus settled, with a degree of finality, disputes as to pathway distribution which have heretofore arisen from anatomical study. Other senses studied by *Girard* are the sense of hearing, sense of touch and muscle senses. Perhaps one of *Girard's* most important findings was that general anesthesia does *not* prevent these electrical responses of sensory nerves, because he was able to measure potential changes in the brain as a result of stimulating a single hair in the skin on the extremities during anesthesia. He thus proved that anesthesia does not inhibit the transmission of impulses to the brain, but merely *inhibits perception* of the impulses, probably by inhibition of synaptic response or depression of the perceptive sense, assuming such to exist as an entity.

[22] Lindsey, Science, 84, 2181, 354.
[23] Gibbs, Science, 85, 2209, 38.
[24] Girard, Scientific Monthly, April 1937.

Girard, further, reports that there are constantly present varying potentials in the brain and nervous system, even in the
absence of the receipt of sensory impulses. The receipt of a
sensory impulse does, however, modify the continual play of these
fluctuations. The nervous system is thus found to be dynamically active and not to be passively awaiting the receipt of sensory impulses as had been previously supposed by physiologists.

Perhaps the most startling finding by *Girard* is that nerve cells
may be affected by activity in other nerve cells in the *complete
absence* of axonic or dendritic connections. Such a process might
be called "teleaction." The conclusion he draws from the foregoing is that the nerve cells do possess a specialized automaticity
which constantly releases impulses of an appreciable frequency
order, along with the already well known reflex quality of nerve
cells, and that under reasonably normal conditions nerve cells are
able to "act spontaneously." Motor cells were found to discharge rhythmically about forty times per second, even under
anesthesia of the sensory supply to the reflex arc. The optic
pathways from the eye to the brain have a rhythmic discharge
rate of about ten impulses per second and this *"rhythmic discharge
completely dominates the whole brain"* says *Girard.* These automatic discharges are not enhanced by light in the eye but are
slowed and even abolished by it, i.e., the rhythm potentials being
present in darkness, but disappear in the light. Obviously, cell
discharge caused by light either interfere with, neutralize, or
modulate this rhythm.

An additional fact is that if sensory impulses are prevented
from reaching the brain by nerve blocking or severance of sensory nerve fibers, the nerve cell rhythm becomes more regular and
possesses a larger amplitude of potential.

Many of the nerve cell discharges are in syntony with other
nerve cells, while some are asynchronous, and unquestionably the
inherent nerve mechanism contributes to these varying rates.
Each cell in the "felt work" of the diencephalon may have its
own frequency, but the rhythm output of the mass is a regular
rhythm of even amplitude. It has been found in the retina that
when all the receptors are stimulated there is a regular rhythmic
discharge in *unison,* but if only a portion of the retina is stimu-

lated, the retinal cells discharge in a chaotic manner and without appreciable rhythm.

The situation, in respect to individual rate discharges by cells and the rhythmic discharge of masses of cells, might well be likened to a symphony orchestra. Each instrument in the orchestra plays its individual part and may possess its own individual rhythm of output. But, under the direction of the leader, the *mass* output of all of the instruments produces a rhythm of output which is easily recognized and is a well known phenomenon observed by the listener.

In reference to the subject of the activity of the cortex and cortical response to incoming stimuli it seems imperative to make clear at this point the fact that the cortex is highly dependent upon an adequate and constant blood supply. If the blood vessels to any part of the cortex be clamped, the part supplied by the clamped artery becomes inactive in a very few minutes. Even the pressure of electrodes on the cortical surface may, by interfering with circulation, render that point inactive, but its power of activity returns shortly after removing the pressure. If for any reason the individual's circulation to the cortex is impaired and lessened, as a result of lowered blood pressure, cervical sympathetic stimulation, or vasodilation of the splanchnic vessels, cortical response gradually decreases. In fact these cortical responses decrease to such an extent that they are no longer an adequate index of the individual's general response conditions when recorded by the oscillograph.

Travis[25] confirmed by *Davis*[26] reports that the oscillographic tracings of brain waves are highly individualized as they pertain to different persons. Some one has suggested that these characteristic brain wave trains made during sleep might be used for identification of individuals, or rather for confirmative identification in conjunction with other well known methods. The individual's brain wave tracings made at different times are so characteristically alike that investigators who have familiarized themselves with one tracing made of the same individual are able to mix the several tracings made by any given individual with a number of other tracings, and can, with a high degree of accuracy,

25 Travis, Science, 85, 2200, 223.
26 Davis, Science, 85, 2209, 8.

select from the mixed tracings those made by the same brain. Since it has been shown that the tracings are not definitely associated with the intelligence quotient of the individual it seems to follow that the inherent pattern action of the automatic nerve cell discharges is sufficiently inherent, and operates under such conditions, so that the output of the "symphony orchestra" response of the automatic nerve cell discharge will be similar in the same individual. An analog would be the ability of a trained ear to detect the difference and to identify by name several musical renditions by the same orchestra.

CHAPTER XI

PHYSICS OF LIGHT

That a clear picture may be had of one of the causes of physiological phenomena later to be discussed, both those determined experimentally, and clinically, a brief study of light energy seems to be imperative. Among the early scientific approaches to the phenomena of light in the photic range were those of *Newton* in 1666. A notable experiment of *Newton* was his dispersion of white light by the utilization of an equilateral glass prism. By this means *Newton* was able to, as he thought, decompose white light into its component colors ranging from red through to violet, and known as the spectral colors. *Newton* assumed as a result of this experiment and his corollary experiment of picking up the spectrum with another prism and causing it to synthesize a beam of white light. Hence his conclusion that white light was composed of the several prismatic colors. He attributed the separation of white light into a spectrum to the differing refrangibility of the several colors present in the white light. It is true that *Newton's* theory did apparently fit the facts as he observed them; however, it is now known that white light consists of trains of irregular pulses of energy in space, and that the wave trains resulting in the so-called colors of the spectrum are actually manufactured within the prism by its action upon the irregular pulses.

This latter statement may at first seem strange to the reader, yet there are numerous instances in physical phenomena where, by an application of some purely physical principle, the result is the transformation of aperiodic energy into a single or many forms of periodic energy depending upon the type of energy and the physical conditions established for its control. That there may be no confusion, *aperiodic* energy may be defined as a form of energy which consists of a continuous flow of an *irregularly* pulsating energy. *Periodic* energy is defined as a form of energy in which the pulsations follow each other at definitely recurring intervals, with like timed inter-spaces between the pulsations. An analog illustrative of this difference can be found in sound. The ordinary noises being aperiodic and consisting of mixed frequencies with mixed amplitudes of energy, whereas the tone produced by a vibrating string or a musical pipe consists of a series of impulses regularly spaced in time and of like amplitude. White light, then, could be considered a "noise," whereas the spectral colors may be likened to the "tone" or pitch produced by a musical instrument. Incidentally, where the word frequency appears hereinafter it will be used as implying a periodic wave train unless specified.

Investigators since *Newton's* time have discovered many facts relative to the nature of light, among them a modification of *Newton's* concept that light consisted of particles or corpuscles which left the source in rectilinear lines, and have evolved the so-called wave theory of light for the explanation of certain phenomena. Recently, while still retaining the wave theory, they have been compelled by other phenomena to attribute to light a form comparable to *Newton's* corpuscular theory. That *Newton* had some idea of this dual nature of light becomes apparent following a perusal of his writings on the subject. To quote *Newton* on this seems apropos at this time.

Nothing more is requisite for putting the Rays of Light into Fits of easy Reflection and easy Transmission than that they be small Bodies, which, by their attractive Powers or some other Force, stir up *Vibrations* in what they act upon, which Vibrations being swifter than the Rays, overtake successively, and agitate them so as by turns to increase and decrease their Velocities and thereby put them into those Fits.

That a clear conception in part, of the nature of light as now understood may be had it seems proper to undertake a consideration of some of the known facts bearing thereon by a review of some of the observed facts of physics and chemistry.

The old conception that elements were made up of atoms which were irreducible into smaller parts or particles has undergone a great modification in the past few years. The old concept that if a bit of chalk crayon be repeatedly broken there would ultimately be reached a point beyond which it could no longer be divided and still remain chalk. If this limit, which is known as the molecule, were to be passed the resultant would be a chemical change and a loss of the attributes of chalk. The disruption of the molecule of chalk chemically would result in one atom of calcium, one atom of carbon, three atoms of oxygen. The former concept held that these resultant atoms were unitary and indivisible. In 1894 *Roentgen*, while experimenting with a high voltage electric discharge tube, discovered that the molecules of gas yet remaining in the tube behaved much as though they would if they consisted of a multiplicity of parts. Further investigation indicated that the atom was composed of a dense central nucleus carrying a positive charge with other bodies rotating around it at extremely high velocities carrying negative charges, in some such manner as our solar system, but not necessarily on one plane of rotation. The smaller bodies carrying negative charges were called *electrons* and were considered to be the units of electrical energy. The electrons were bound to the positively charged nucleus by reason of their opposite charges, but due to their *like* charges the electrons themselves never make contact each with the other and are thus caused to maintain a definite space relationship, within the limitations of attraction by the nucleus and their own mutually repulsive forces. The changes in space relationship, number of electrons and their grouping about the nucleus are the distinctive differences between the atoms of the several elements.

It was early noted that the total negative charge possessed by all of the electrons in any given atom exactly balanced or neutralized the positive charge on its nucleus, despite a mass difference between the nucleus and the electron, a ratio of 1840 : 1. By reason of the neutralization of charges on the nucleus and the total of electronic charges, each by the other, the spatial charge of the unmodified atom is, of course, zero.

Under a number of types of stresses, electrical, heat, chemical or radiant energy, it is possible for an atom to lose an electron from one of its orbits, usually an electron having high energy content in an outer orbit. The result of this loss of an electron is an entity now carrying a relative positive charge, due to a loss of a part of its neutralizing negative charge. This entity is called an *ion*, more properly a positive ion in this instance. Conversely, a *neutral* atom might have attached to it a free electron by picking up one in its immediate vicinity, because a neutral charge is relatively positive to the negative charge carried by the electron. Immediately, upon this attachment the atom loses its status of neutrality and, because it now carries an extra negative charge, is called an ion, in this instance a negative ion.

Upon these phenomena of gain of an electron or loss of an electron—ionization—depend all known chemical reactions, electrical reactions, nerve transmission, and the processes of metabolism of the living unit. If, by some process, all ionization could be nullified, life and function as we know them would cease.

The electrical nature of the components of the atom is easily demonstrated in the laboratory by producing a gaseous electrical discharge, then by applying a magnet along the side of the tube, noting the deflection of the stream of ions, and perhaps some free electrons, in the path of the discharge.

There are many forms of radiant energy which may be caused to leave a source, yet which act at a distance, light being the common example. How this action at a distance takes place was difficult to conceive until the ether hypothesis was evolved. The ether was assumed to be an ever present structureless, universally elastic substance which pervades all space and matter. *Newton*'s original corpuscular theory did not require the assumption of an ether for light propagation, but when *Huygens, Young,* and others developed their wave theory of light propagation, it became necessary to ''devise'' a medium through which waves could be propagated. Whether the ether, as originally hypothesized, exists is at present a moot question, despite the fact that scientists are in agreement that the several forms of radiant energy are fundamentally wave-like in character.

Clerk-Maxwell classified these several forms of wave energy, on a frequency basis, into his now universally accepted electromagnetic spectrum. Starting with a direct current of one cycle of infinite length, by a process of halving wave length, or doubling the frequency, divided the spectrum into units known as octaves. At present there is known to be a total of sixty-two octaves, with perhaps more to be discovered, in the electro-magnetic spectrum in the order from direct current, through the alternating currents, long Hertzian waves, short Hertzian waves, infra-red, the photic range, commonly known as the visible range, ultra-violet range, Grenz rays, soft and hard x-ray, through the radiations of radium to the cosmic rays of Millikan. All of these energies, regardless of frequency, have a constant velocity of three hundred million meters per second, something over one hundred eighty-six thousand miles per second. All of these forms of energy travel in the form of wave-like disturbances, or set up wave-like disturbances, in the hypothesized ether.

In the photic range use is made of the wave theory by opticians to determine curvatures of lenses, mirrors, setting up of interference, polarization, and the other easily demonstrated properties of light. While it is true that from a purely optical standpoint the wave theory seems to be all-sufficient for the calculation of optical instruments and devices, there do exist certain properties of light in the photic range which cannot be accounted for by an application of the wave theory. For instance, the astronomer does not hesitate to reflect light emitted by a star from the surface of his concave mirror, catch the converging rays by a reflecting prism, pass the pencil of light outside its entrance path, interpose a convex lens, and bring the rays of light to a focus on a photographic plate. He uses and applies the wave theory from the star to the concave mirror, from the concave mirror to the reflecting prism, from the reflecting prism to the convex lens, and from the convex lens to the surface of the photographic plate. All of these phenomena down to the surface of the photographic plate can be correctly anticipated and calculated by an application of the wave theory, but he cannot thus account for what takes place in the photographic plate following the impingement of light thereon.

The sensitive surface of the photographic plate consists of a thin layer of gelatin containing one of the halides of silver, usually the bromide. This salt of silver possesses the property of being so changed after exposure to light that if it be placed in certain chemical solutions, subsequent to the exposure to light, there will be precipitated into the gelatin small masses of pure metallic silver. By the arrangement and density of the clumps of bits of silver, in the vicinity of the points where the light was focused by the convex lens is produced a negative image of the star sought to be photographed by the astronomer. By *negative image* is meant that the picture thus produced will be black where the intensity of the light was greatest, and some shade of gray where the intensity was less, and clear where no light was focused.

Obviously, the change produced by the light in the bromide of silver was some form of a chemical change, because those particles of silver bromide that were not exposed to the light, and which lay immediately adjacent to those upon which light fell, did not undergo the phenomenon of precipitation of metallic silver when placed in the chemical solution. In other words, light possesses some of the characteristics of other well known chemicals. It probably would seem strange to walk into a chemist's shop and request a quantity of light by the gram or pound as one might purchase other chemicals, yet the fact remains that light carries chemical potentialities just as do other chemicals which may be purchased by weight.

This chemical property of light confused physicists for a long time. This property could not be accounted for on the basis of the wave theory, because, as has been said above, electrons possess some mass, although small, and it is inconceivable that they could be caused to leave an atom by the mere impact of a massless wave. Incidentally, waves possess no mass except as mass is an attribute of the media through which they move. The loss of an electron under the impact of light, as we have seen, is a chemical change, and a chemical change either absorbs or releases energy, which latter phenomenon can only be observed by its action upon mass.

If the chemical change is not a function of the wave, then of what is it a function or a result? The solution ultimately determined upon is that light is not wholly composed of waves and they *must possess* something comparable to mass if they are to be capable of moving mass, even so small a mass as an electron. Investigation shows that when light falls upon polished surfaces, or in fact upon any surface, electrons leave that surface constantly for the duration of its exposure to light. Furthermore electrons so released take the same direction as the reflected ray if the surface be sufficiently polished to result in specular reflection. The light leaves the reflecting surface with a velocity equal to that of the light incident upon the surface, i.e., three hundred million meters per second. But the electrons leaving the surface at the same time under the influence of light do not have any such high velocity. Furthermore, the velocity of the electrons leaving the surface varies, and this variation in velocity is directly proportional to the frequency of the incident light.

Thus with high frequency light incident upon the surface there will be emitted high velocity electrons, and with low frequency light the electrons emitted will have a much lower velocity. The ability of an electron in motion to do work is a function of its velocity squared, which means that a mere doubling of the velocity quadruples the work doing power of the mass. *Sperti*, then at the University of Cincinnati, long ago demonstrated that electrons are *released from atoms which enter into the composition of living cells and tissues,* in like manner as they are emitted from inanimate material under test conditions in the laboratory. The significance of this finding is that the chemical structure of the cell can be caused to vary by the light incident upon and absorbed by it, and, what is more, the degree of chemical change becomes a function of the frequency of the absorbed light. It should be kept in mind here that the velocity of the light does not change under the conditions under discussion, but that only the velocity of the electrons emitted by it may change and that this velocity is purely a function of the frequency of the light.

Intensity of light falling upon matter in nowise alters any of the above. Intensity does not cause a variation of electron velocity. Intensity does, however, determine the number of electrons released, but the work done by each electron is determined by its velocity, which in turn was determined by the light frequency. If quantity response of electron emission is desired, obviously an increase in the intensity is a desideratum, while if chemical change is the desired end-result, frequency variation is the method to be employed.

Another fact relative to the emission of electrons under the influence of light, which plays an important part later in this discussion, is that the release of electrons upon the impingement and absorption of light is instantaneous, there being no measurable lag between the time the light falls upon the surface and the appearance of electrons in space. In anticipation of what is to be said later it should here be pointed out that it is a good thing for the animal kingdom that there is no lag between the time the incident light enters the eye and the beginning of the photochemical change which is to result in a sense perception being transmitted to the brain. If such a lag did exist there would be numerous instances in which the animal would not have sufficient warning of danger to protect itself.

If the waves do not release electrons the question logically arises, what does release them? As has been indicated above mass or something comparable to mass seems necessary. In order to get a fairly clear picture of this mechanism it is necessary to go back to the fundamental picture of the atom. Here we find electrons rotating around the nucleus in orbits of differing radii. Electrons in orbits of shorter radii possess a lower energy content than those rotating in orbits of the longest radii in that atom. When energy is absorbed by the atom it expands, which is just another way of saying that the several radii of the orbits increase in length. Obviously, it is possible that by absorbing energy a point would be reached by electrons in the outer orbits, the one having the longest radii, where their attraction for the nucleus would be *less* than the centrifugal force created by rotation. If and when such a state is reached the electron will leave its orbit, flying off at a tangent to the radius at the point of its release. Such a happening would be *ionization* and would leave a positively charged ion. Also, such a happening would constitute a chemical change. An instance of such a happening and change is that which takes place upon striking a match. The mechanical energy produced by the friction of the inflammable head of the match upon the friction surface being sufficient to remove electrons from the surface atoms. These ionized atoms being in a nascent state combine with free oxygen in the air and produce combustion. This combustion emits light. This is one method of producing light and was the common method in use until comparatively recent times. Light can also be produced by chemical change brought about by an electrical current, as is found in the electric arc. Both of these, remember, result in the ultimate destruction of the solid during the chemical changes, of course, allowing for the fact that there is no complete destruction of matter, it merely being a change in form. The two foregoing sources of light are atomic sources of light in that the atoms enter into new chemical combinations during the processes attendant upon the emission of light. There is yet another form of atomic light emission in which atoms are excited, but in which there is little if any chemical change taking place, such as the commonly seen glow discharge tubes used in signs and for decorative purposes.

In each of the three foregoing types of atomic emissions of light, light is emitted at the instant a free electron falls into the orbit just vacated by the electron emitted as a result of increased energy content of the atom. At the instant this electron falls onto the orbit a unit quantity of light is emitted, which represents the exact amount of energy absorbed by the emitted electron and which energy caused it to leave its orbit. This unitary amount of light energy has been given the name *quantum,* but is also known as *photon.*

There is another method of causing the emission of light in which no chemical change takes place. This is the incandescent source of light in which some element or compound is heated to a high temperature in the presence of inert substances which prevent chemical change. Both gasses and solids may be raised to incandescence by increasing their temperature. The mechanism of the emission of light by incandescent sources differs somewhat from that described for atomic sources of light. The assigned mechanism roughly stated is this: When energy is added to the atom by absorption of some external force the electron moves away from its nucleus in the path around its orbit, thus increasing its radius. If the energy absorbed is not enough to drive it completely out of the atom, the electron eventually drops back to its original orbit, an orbit of shorter radius than the "expanded" one. When the electron falls from this hypothetical orbit of longer radius back to its original orbit, an orbit of lower energy content, the excess energy which carried it into the outer orbit is now released into space as a unit of light. As can readily be seen the electron describes an undulating path instead of a circular path around its orbit. At each drop of the electron in undulation to its original orbit radius there is emitted a single quantum of light, the number of undulations during one circuit around the atom, with their multiple emissions of quanta, determines the frequency of emitted light. Consequently, the higher the temperature the greater the number of undulations and the higher the frequency of emission. Bear in mind that it requires an energy addition to increase the length of the radii of electrons in the atom, and that this *same amount* of energy is released when an electron falls from an orbit of higher energy content to one of lower energy content, i.e., falls to an orbit nearer the nucleus.

The unit of energy possessed by each quantum is calculated by the formula $E_q = Fh$, where F is the frequency of light, and h is Planck's constant, which is 6.54×10^{-27} ergs per second. Obviously, this is a very minute amount of energy and in all probability could not be detected if only one quantum were emitted. But considered in terms of quantitative emission we find that the sun emits about 1.98 calories per minute per cm.^{2}, which figure when applied to a larger area amounts to slightly more than five million horse power per mile^2. This is indeed a lot of power. Because this power factor will be considered later, let it be understood here that the area of the pupil of the human eye is roughly .5 cm.^{2}, which means that in clear sunlight it absorbs approximately one calorie per minute.

That there may be no confusion between electrons and quanta it should be noted that the formula for calculation of the work done by the electron is identical with the formula for the calculation of work done by the quantum. With this in mind the following mathematical formula can be written:

$$W_q = Fh$$
$$W_e = Fh$$
$$\therefore \quad W_q = W_e$$

But it is known that the electron possesses mass, hence the only logical conclusion that can be drawn from the above formula is that the quantum possesses mass or something comparable to mass. If it possesses true mass it should be subject to the laws of gravitation just as are other masses. The foregoing conclusion is one of the early Einstein hypotheses, which astronomers have sought to prove or disprove by efforts to measure the deviation of the stream of light coming from a far distant source as it passes larger bodies in space, such as the sun. Should such a deviation be measured, it would rightly be considered proof of the foregoing conclusion by Einstein. Let it be said, however, that there is still some debate as to the data secured in these experiments in efforts to measure gravitational deviation of light by astronomers.

Almost any atom can be caused to emit light by adding energy to it, but the light emitted by the atoms of any given element always possess the same frequency, i.e., sodium always emits the two "D." lines in the yellow region of the visible spectrum.

Other elements, of course, emit other lines, but they are never the same for any two elements. Also, the excited atoms of any given element always absorb the same frequency of light that they would emit if in a state of excitation, but not in the path of incident light, this being a completely reciprocal relationship.

Some radiators are, of course, better than others due to their atomic structure and the relative looseness of the arrangement of the nucleus and electrons. The best radiators are also the best absorbers of energy.

In the laboratory the best radiator or absorber of radiant energy is a black body, which may take either of two forms: A perfect black body being merely a hole, which, incidentally, is capable of radiating and absorbing energy simultaneously; an imperfect black body would be any blackened surface made by the application of a pigment, or resulting from an inherent quality of the material itself. Such a black body may either radiate or absorb, never both simultaneously. The human eye possesses an example of each of these two types of black bodies: the pupil being a perfect black body which may absorb and radiate energy simultaneously, which fact is put to practical use by makers of optical instruments for the observation of the interior of the eye. In this instance light is passed into the eye by the device, while light reflected from the interior of the eye passes out through the same pupil for inspection by the observer. The other black body possessed by the human eye is one of the imperfect type, the choroid. This black body possesses a pigment which absorbs all frequencies of light, in fact all light, which passes through the retina and falls upon the choroid. Light so absorbed is probably transformed into heat which is carried away by convection by the highly vascular bed of the choroid.

As will be mentioned later, there is another possible effect in the choroid due to potential differences which may exist between it and the retina. This possible effect will be discussed theoretically later, and in view of certain well known physiological phenomena.

We now have before us two concepts of light, the wave theory and the unitary or quantum theory. The quantum theory holds light to be somewhat like a string of machine-gun bullets following closely behind each other. In fact the present concept is comparable to *Newton's* original corpuscular theory. It seemed impossible to reconcile these two theories and yet they seemed absolutely necessary to account for the types and kinds of phenomena exhibited by light. *De Broglie*, a French scientist, advanced his "waveicular" theory, in which he cited the fact that all moving particles always set up waves in the medium through which they move. The moving boat sets up waves in the medium through which it moves; the moving bullet sets up waves in the air through which it moves. *De Broglie* hypothesized that once these waves were set up, their reaction upon the particles would determine the direction they would take. Also, that any influence upon the wave would alter the direction of the particle. Both of these hypothetical assumptions have been repeatedly confirmed. In an effort to explain the *De Broglie* theory to upper classmen, *Sir J. J. Thompson* of Oxford, used the following analog: The gossamer spider resting upon a limb of a tree might wish to change his position or to move to a distant point. To do this he spins in the air a few filaments of web, and with the first gust of wind releases his hold on the limb and is launched into space attached to what constitutes a small parachute. Once free in space, the spider's motion is now controlled and dependent upon the web filaments which support him, corresponding to the waves set up by the free quanta or free electrons in space. Obviously, any interference with the free movements of the web filaments, such as by a twig, a weed leaf or other object, would alter the direction of the moving "quantum," i.e., the spider, may cause it to be reflected or refracted, or may produce other well known optical phenomena.

All free quanta and free electrons possess these waves. Both react in the same way as particles and are guided by the waves they have set up as soon as they are free in space.

The "power factor" of the electron varies with its velocity which as has been said heretofore, is a function of the frequency of the light which frees it. Therefore, the ability of the electron to do work varies through wide limits. Electrons set free by the visible range of the spectrum can only vary in power from unity to approximately four times unity, because the frequency of light in the photic range just practically doubles throughout that range, which is almost an octave of the electro-magnetic spectrum, since work done varies as the square of the velocity. The velocity of the quantum, however, does not vary while it is transversing the same medium. This means that the wave-length of its attendant wave remains a constant in a given, or the same medium. But, should the quantum enter a medium which retards its velocity the wavelength shortens, while the *frequency remains constant*. A practical application of this fact and the foregoing facts will be considered later.

Another physical phenomenon of light of which use will be made later in this presentation is that light waves of low frequency are dispersed less by prisms or lenses than are light waves by high frequency. The phenomenon is usually known by the phrase "chromatic aberration." As a result of the difference in dispersion between the two extremes of the spectrum it is found that red light is brought to a focus at a point farther from the nodal point of a lens than is blue light. This is particularly true in the human eye and has been shown to constitute a focal interval of about 1.8 D. between the red and the blue for a light source at one-third of a meter from the nodal point of the eye.

This phenomenon can also be used clinically by taking advantage of "edge diffraction" when using a bichromatic source, such as the transmission of cobalt glass, and has a practical application in the use of the syntonic technic.

REFERENCES

Bohr—Atomic Theory.
Lemon—Galileo to Cosmic Rays.
Zworykin—Photocells.
Wood—"Physical Optics."
Heyl—"New Frontiers of Physics."
Johnson—spectra.
Reiche—"The Quantum Theory."
Haas—"The World of Atoms."

PART III

CHAPTER XII

OCULAR PHENOMENA IN RESPONSE TO LIGHT

In the lower orders of living things there is a wide variety of photoreceptors. In some instances these take the form merely of a few cells which, due to their pigment, are capable of absorbing a greater amount of light energy than other cells in the same structure. In more organized living things there are larger groups of cells which seem to function simultaneously as a group upon impingement by light. In the more complex organisms there is usually some special form of sense organ containing highly specialized cells which respond only to light. In most of the animals this layer of specialized cells is found in the retina, and, in the highest orders, consist of two types of cells known as rods and cones, the photoreceptors. Figure VIII is a diagrammatic representation of a cross section of a cone. Indicated on its external surface and internal surface are the electrical potential difference symbols, in terms of the polarity, that exists between the inside of the cell and its outside. Also shown is the nerve fiber which leads from the photoreceptors to its synapse with a bipolar cell in the body of the retina.

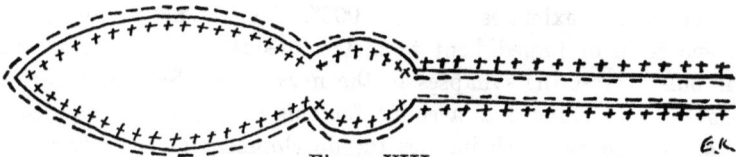

Figure VIII

Light incident upon the cone sets up a wave of depolarization in the photoreceptor, which, when it reaches the nerve fiber, normally increases polarization at that point. This increased polarization starts a nerve impulse down the fiber enroute to the brain, which route includes several synapses, two synapses at the bipolar cells, also, at the ganglionic layer of the retina, in the thalamus, where there may be a multiplicity of synapses, and finally in the cortex itself. Figure IX is a schematic diagram, in cross section, of the retina showing the photoreceptors, axones, bipolar cells, and the ganglionic layer.

Continuous incidence of light on the photoreceptors elicits a series of impulses in the optic nerve tracts, thalamus and finally in the cortex. The interpretation of these impulses constitutes the conscious act of vision.

Enroute from the retina the impulses traveling down the optic nerve fibers must pass a number of synapses. Synapses have several properties which enter into and to some extent alter the transmission of impulses across them. The first and most important property is that of "synaptic delay." This delay at the

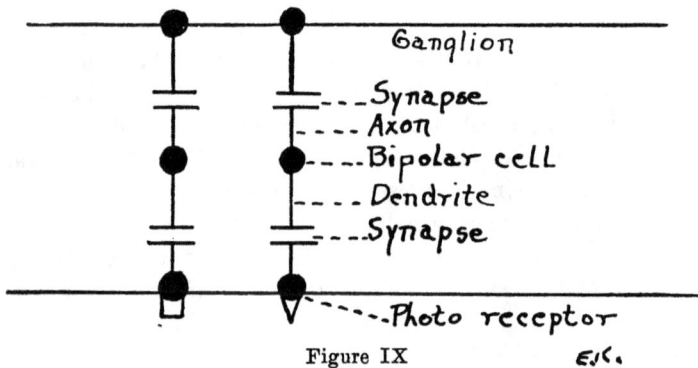

Figure IX

synapse approximates two σ, .002″. This synaptic delay is some form of impediment to instantaneous transmission of an impulse across the synapses to the nerve cell. Several theories have been advanced to account for this delay: 1. The so-called "soup" theory which implies the development of some form of chemical mediator between the dendrites of the synapses, which, when it reaches a certain concentration following the receipt of a series of impulses at the synapses, so lowers the resistance across the synapse that the impulse may stimulate the dendrites of the next cell. Obviously, this requires an appreciable lapse of time, and, as we have said above, approximates two σ.

Another theory of synaptic delay is the so-called "All-or-none" theory. This theory requires the assumption that the dendritic fibers become so fine and are of such high resistance that they cause the impulses to take a longer time to pass through them to the dendrites of the cell in the synapse. Such fibers would have to be sub-microscopic in size if they exist at all. More recent investigation tends to indicate that this theory is a less tenable one than the chemical mediator theory.

2. Another peculiarity of the synapse is that of the so-called "after discharge," a kind of repetition of the impulse in the absence of a further impulse stimulus.

3. Another peculiarity is the unidirectional transmission of impulses through the synapse. It is a well known fact that stimulation of a nerve fiber any place along its path results in transmission of an impulse in both directions along the fiber from the point to which the stimulus was applied. Most fibers enter a synapse at each end, but these synapses have a sort of "check-valve action" by means of which impulses over sensory fibers can only pass toward the brain, while those in motor fibers can only pass toward the periphery. So far no adequate theory has been developed to account for this one way action.

4. Another peculiarity of the synapses is that known as temporal summation. This effect results when a stimulus is of too low an intensity to excite a response when applied as a single stimulus, but succeeds in exciting a response and passing a synapse if it be repeated, provided the repetitions are close enough together in time. If the several stimuli have too great a time lapse between them there is no summation effect, probably due to the evanescence of any chemical mediator which may have resulted from the prior stimulus.

5. Still another peculiarity of the synapse is that of spatial summation, which results when two or more nerve fibers reach the same synapse, both of which are stimulated. If one alone of these fibers be stimulated it is usually impossible to cause a response but if more than one is stimulated we have the phenomenon of re-enforcement of each other in an additive manner. It has also been found that adjacent fibers in the same nerve trunk may exhibit a re-enforcing effect upon each other even before their impulses reach the synapses, but that if any fiber of the group should enter a different synapse, this fiber will excite a response when many of the fibers of the nerve are stimulated, whereas it might not excite a response if this single fiber alone were stimulated.

6. Another synaptic phenomenon is that of inhibition by the synapse. This may be due to interference between paths if there

should be a greater intensity of stimulation in a different adjacent path. More recent physiologic experimentation indicates that there is a possibility, even a high probability, that this inhibitory effect may result from a neutralization of an impulse by an out-of-phase impulse in an adjoining path or an out-of-phase automatic discharge by the nerve cell of the synapse. Some investigators have hypothesized the existence of inhibitory cells, while others have held to the belief that there may be horizontal cells having parallel function, neither of which contentions seem to be in agreement with the now known automaticity of rhythm discharge by nerve cells. In a former chapter the theory was advanced of a standing positive charge on the retina as a whole, and at that time certain facts were presented in support thereof. While this theory may not stand future critical investigation the fact remains that applications of suitable stimuli, made with this theory in mind, show a very definite change in body potential, also an altered respiratory rate and in most instances an altered pulse rate. All of these changes are produced by the mere act of applying a proper stimulus to the retina which meets the conditions necessary for altering this theoretical positive standing charge.

The so-called duplicity theory of vision is based on the fact that there are two types of receptors in the retina, a photoptic mechanism with a special receptor which functions during relatively high intensity of light, and a scotoptic receptor and its mechanism which apparently only functions at low intensity illumination. The former receptor is accepted to be the cone in the retina, whereas the latter is attributed to rod function. At high intensity under the cone function it is possible to perceive and differentiate color hues, while under the scotoptic function, where rods alone are apparently functioning, there is no color perception, such vision being practically achromatic.

Microscopic examination of the retina establishes the fact that the foveal region, consisting of cones only subtends an angle of about one-half of a degree with the nodal point. Furthermore, there is a single nerve fiber from each one of the cones, which fibers pass into the optic nerve trunk. Histological examination of the retina, as we move from the foveal region toward its periphery, shows an inter-mingling of rods with the cones and the farther removed the observed point is from the foveal region, the greater is the ratio of rods to the cones. In fact at the extreme periphery of the retina there are no cones to be found microscopically.

Another significant fact derived from such an examination of the retina is that a multiplicity of rods, often as many as sixty to a hundred, are connected to a *single common nerve* fiber which passes into the optic nerve. It follows from this fact, taken in connection with the single-cone-single-fiber finding, that vision in the periphery is much less accurate in regard to form perception than at the foveal region. On the other hand, although it functions achromatically, the peripheral region is much more sensitive to illumination, probably due to the multiplicity of rods sending impulses jointly to the same nerve fiber.

If a test light spot of high intensity is thrown upon the retina at different distances from the fovea it is found that the image appears brightest when focused in the high concentration of cones in the fovea, but that it appears to be dimmer or of least intensity when focused on the periphery. Such a finding seems to be anomalous in view of what has been said above until it is recalled that we are dealing with the eye in its photoptic or light adapted state, in which condition the visual purple in the rods is virtually bleached causing them to be less sensitive. The finding above mentioned follows in rather close agreement the distribution of cones across the retina from the fovea toward the periphery.

Contra if the same experiment be repeated with a test-light object of *low* intensity and the measurements made in a dark adapted eye, made scotoptic by keeping the eyes in darkness for not less than fifteen to thirty minutes, it is found that sensitivity is lost at the fovea and may even reach zero in that region, but that the apparent intensity is increased in the perimacular region with a slight falling off at the extreme periphery.

The sensitivity of the retina for color also varies as we cross the retina from the fovea toward the periphery. The sensitivity for color does not follow the spectral order of red, orange, green, blue, violet as would be expected if the demonstrable phenomena were caused by chromatic aberration or dispersion by the lenticular elements of the eye. Leaving the fovea, as the center, it is found that the field for the perception of green is the smallest in the extra foveal region; next in order we find the field for red; and the largest field for color perception is that for the blue. This seems easy to understand in view of the work of *Chase,* cited *infra,* whereby was demonstrated the existence of a blue sensitive substance in rhodopsin which is present in the rods. Incidentally, the size of the color field, i.e., ability to differentiate hue, depends somewhat upon the size and the brightness of the test object, but, regardless of this size or brightness, the field relationship remains the same, moving from the fovea, and is in the order of green, red, blue.

The ability to recognize form and to make minute discriminations varies markedly between the foveal region and the perimacular and peripheral regions. One reason for this seems obviously to be the association of single cones with single fibers, so that an image falling upon the multiplicity of cones would stimulate them variously and each fiber would carry into the brain an impulse frequency determined by the intensity of illumination or lack of it, produced by that part of the image which fell upon its cone in the foveal region. The ratio of cone to fiber of $1:1$ is, therefore, important for fine discrimination of detail as also is the relative size of the cones. By analogy we might cite the difference between the appearance of a half-tone of sixty lines to the inch in a newspaper, and a half-tone of the same scene on a highly calendered paper, but containing two hundred lines to the inch. The coarser the screen, the greater the loss in detail, and the finer the screen the greater the ability to discriminate detail. Furthermore, in the coarse screen the fine qualities of shading discrimination are impossible as contrasted with the fine screen. It follows, then, that the histological structure of the foveal region, in terms of size of cone, has an important bearing upon detail discrimination as well as the intermediate tones. The peripheral region with its ratio of $60:1$ or $100:1$ cones per fiber, could not, therefore, lend itself to high detail discrimination nor to high degrees of differentiation of shadow or shadow density.

The rate of regeneration of photo-substances in the photoptic eye has been found to be relatively high in that it is relatively complete after two or three minutes of darkness. On the other

hand, the rate of regeneration of the photo-substance in the scotoptic eye, in the rods, takes a much longer period of time and is
not complete until about thirty minutes of dark adaptation. We
can, therefore, consider that in the light adapted eye the rods are
practically out of action and that light is perceived only by the
cones, the same being responsible for fine discriminations.

Sight must not be lost of the fact that there is a constant variation between the two states of seeing—photoptic and scotoptic.
Co-existent with the photoptic vision there is also some response
by the rods to some intensities of light and vice versa. The
process of dark adaptation cannot, however, be limited to a change
of vision from cones to rods, nor, conversely, can light adaptation
be restricted to a change of vision from rods to cones. It must
constantly be kept in mind that these two states, while they function separately, also to some extent function simultaneously.

The ability to achieve a high degree of visual discrimination,
being a cone function, can be understood when it is stated that
there are approximately a hundred and fifty thousand cones per
square millimeter in the foveal region, and that each of these
cones only sustend an arc of about forty seconds at the nodal
point. When an image with a sharp line of demarcation between
light and dark falls on this region, the cones that are just outside
the dark region will initiate higher frequency responses than
those in the dark region, which latter may not initiate any response at all. When an image falls on the fovea in which the
intensity gradually varies from darkness to light, those cones
occupying an intermediate position between the extremes of dark
and light will transmit nerve impulse frequencies of some intermediate degree of frequency between zero impulses and the maximum of which the cone fibers are capable. This will result in the
perception of half tones of varying degrees of density. During
the act of reading, where only extremes of dark and light are
present, the rate at which the images fall upon the foveal region
is extremely high, particularly when it is remembered that there
is a constant motion of the visual axis, somewhat like nystagmus,
in an arc of about one or two minutes. This lateral movement has
a time period that is rather high, but still not high enough to
prevent the average five-letter word being read in about one-tenth
of a second. This means that due to the lateral twittering motion

of the eye, taken in conjunction with the impinging of the dark area in the fovea, the cone receptor must receive and transmit an impression in roughly one-two-hundred-fiftieth of a second. It can be seen that the range of frequency that the cone and its fiber must handle is high and that the degree of accuracy of perception is considerably greater than is commonly supposed. It follows that good visual acuity is dependent upon both the small size of the receptor and the resolving power of the lenticular system of the eye, and yet it is also dependent upon the time constants mentioned above and the transmission of the nerve impulses at the frequencies required.

The following point in reference to light intensity and visibility will be discussed more at length at a later place in this chapter, but should be mentioned here in order to prepare the reader for its more detailed elucidation. It has been found that the larger the area of light patch stimulus, the lower is the degree of brightness necessary to permit visibility. It follows, therefore, that there is a partially reciprocal relationship between area and intensity.

The extremes of intensity under which the retina as a whole can operate are found to be from 10^{-8} to 10^{-4} foot-candles per square foot. The range of cone perception of intensity is from 10^{-4} to 10^{4} foot-candles per square foot; whereas the range in the rods is from 10^{-8} to approximately 10^{-1} foot-candles per square foot. These ranges are the extremes and show the greater sensitivity of the photoreceptor substance in the rods to that of the cones.

Perception of movement of objects in space is dependent upon movement of the image over the retina, with its consequent change in distribution of light thereon. It seems that some parts of the retina are more sensitive to *changes* of stimulation than others. It follows that such parts would more easily detect motion than those not so sensitive, assuming the degree of illumination and the light adaptation state to be the same. In the study of flicker phenomenon it has been found that the peripheral region shows a very marked decrease in time for flicker perception over that of the light adapted foveal region. The sensitivity to flicker depends upon the intensity of the light field and consequently upon the degree of adaptataion at the time.

Following dark adaptation in faint light it is found that the luminosity curve is widely different from that in the photoptic state. In the photoptic state the maximum sensitivity is in the yellow-greens, while in the scotoptic state the curve has its maximum scope in the blue-greens. It seems to follow that during partial dark adaptation at night the use of red signal lights or pink absorption tints in lenticular corrections are not physiologically proper procedures, assuming an equal intensity of light. Restated, it may be said that in the scotoptic state the eye is much less sensitive to reds and yellows, and more sensitive to the blue-greens. Under conditions of fog and haze, since the low frequency reds are more transmissible than the higher frequencies, it appears that there may be some advantage of using these frequencies, but in the absence of haze they should be avoided.

Miners and others working for long periods under low intensity of illumination often develop a nystagmus, a lateral oscillation of the eyes in their orbits. This oscillation results in shifting the image of the retina over adjacent cones, thus bringing more cones under the influence of low illumination image. This, in all probability, is nature's method of enhancing the vision of the subject under low intensity illumination, due to the fibrous summation effect which will be discussed later.

It has been shown that the sensitivity of the retina is lowered quite rapidly when a fairly intense beam of light enters the eye, which loss of sensitivity is far greater than the rate of photochemical changes which take place. This finding is in agreement with the well known observation that the apparent brightness of an object is in part determined by the intensity of the light which surrounds it. It follows that any study of the relationship between acuity and the intensity of illumination on the object of regard is only of significance when the intensity of illumination of the surroundings are also specified. The sensitivity of the eye to flicker is increased by increasing the intensity of the light in its surrounding. The ocular hygienic conclusion to be drawn from the foregoing is that any system of illumination which provides high intensity in one position, without at the same time increasing the illumination of all of its surroundings, is bad and results in lessening acuity with consequent eyestrain. If the surrounding illumination be adequate, the efficiency of the eye to discriminate small objects continues to improve with increasing levels of illumination up to about a thousand foot-candles, says *Wright, infra.* Good ocular hygiene, then, requires a high intensity of illumination in an entire room, but for fine detail work an additional intensity should be cast upon the work in hand.

Styles, quoted by *Wright* observed the rather unexpected phenomenon that the apparent brightness of a patch of light depends among other things on the angle at which its beam is incident on the retina. The lower the angle the higher the apparent intensity. Experimentally this can be shown by the use of two narrow concentric beams of light, one which is so narrow that it passes through the center of the pupillary area, and the other having the diameter of the pupil itself. If the two sources of light have equal physical intensity, the beam passing through the center of the pupil will appear to be much brighter than the one occupying the entire pupillary area, despite the fact that a larger based cone of light is concentrated onto the focal point. *Styles* suggested that this marked difference in apparent intensity might result because the more oblique rays must necessarily travel through a greater thickness of pigment due to their angularity. *Wright* disagrees with this and prefers the explanation that light must pass directly along the axis of a cone to produce its maximum effect. To this writer the latter seems to be the more logical conclusion of the two mentioned.

That the photosensitive substance in the retina has the power of self-regeneration was demonstrated by *Hecht*[1] who reproved an older finding by *Kuhne* to the effect that solutions of rhodopsin bleached by action of light had the power of self-regeneration to an approximation of its original saturation density when kept in darkness for a proper time. *Chase*[2] finds that there is apparently one or more substances involved in this response. He found that when solutions of rhodopsin were bleached under the influence of a photographic flood lamp, which is merely an over-volted tungsten filament light bulb, that the bleaching was greater and the subsequent regeneration of color was to a greater degree than when the rhodopsin solution was bleached by the ordinary 100

[1] Hecht, Science, 84, 331.
[2] Chase, ''Accessory Photosensitive Substance in Visual Purple Regeneration.''

watt lamp being supplied with its rated voltage. At first thought this increased bleaching response might be accounted for on the basis of the higher frequency emission, more of the blue rays, by the photo flood lamp resulting from its over-volting. As has herein before been said, the energy content of the higher frequency lights is greater than that of the lower frequencies, with a consequently greater photophysical and photochemical change.

While the foregoing well known physical facts account for the increased rate of bleaching they do *not* account for the increased rate of regeneration of the rhodopsin previously bleached by the photo flood light. Apparently the only way to account for this phenomenon is to assume the existence of a blue sensitive fraction in the rhodopsin, whose decomposition was essential for the regeneration of the visual purple. On the basis of this assumption, solutions of visual purple bleached under violet or blue light should show a rate of regeneration *greater* than that of a solution of the visual purple bleached under green, yellow or orange light. Experimentally, this proves to be true as is shown by the following experiment. Passing a blue-white light from a photo flood lamp through a yellow filter, Corning 350, and using the same light source passing the light through a blue filter, Corning 554, it is possible to isolate these regions. In the former instance a blue minus light will be transmitted, while in the latter a virtually red minus light will result. If a previously prepared solution of rhodopsin which has had its pH controlled to about 7.7, on the alkaline side, is divided into two portions and separately illuminated, the one by the blue light, at a distance of eight to twelve centimeters, for thirty minutes, a time three hundred percent. longer than necessary to secure complete bleaching, it was found that the rate of regeneration of the visual purple in a five millimeter absorption cell, measured at 500 mμ in darkness for the succeeding thirty minutes showed a density increase by about .0330. This photometric density being calculated as log $l_o/1$. The other specimen was irradiated with the yellow light and measuring, its increase in density over the same period of time and under the same observation condition, i.e., in darkness at 500 mμ, showed a density increase of only .0037.

As a check on this experiment to learn if the yellow bleached solution was capable of a further regeneration in visual purple, it was now illuminated by the blue light for thirty minutes until it completely bleached. Measurements were then made under the standardized conditions mentioned above and it was found that there was a density increase of .0330, which is the identical figure that was secured by the single irradiation under the blue light of the other half of the specimen.

Chase also measured the density decrease at 450 mμ and found that the sample bleached by the yellow light was decreased materially in density during only a ten minute exposure to blue light. He draws the conclusion that the marked regeneration at 500 mμ as well as that at 450 mμ occurred subsequently to the density decrease in a blue absorbing substance in the rhodopsin.

Just what part a blue decomposing substance in visual purple may play on the sense of vision is not now known, but it does seem to be an important fraction in the regeneration of visual purple in darkness, i.e., the production of a scotoptic retina.

The findings of *Chase* are in agreement with those of *Wright*[3] who says that the function of the retina on which light falls, whether it be foveal or extra-foveal, is governed by the previous stimulation of the portion of the retina being used. Continuing he says: ''Again, the sensation will be very much affected by the colour and intensity of the illuminated area surrounding the test spot.''

In an effort to attain a more correct concept of the *effect* of stimulation of the retina by light a number of experiments have been made. One of the first steps in these experimentations is to expose the nervous elements involved in seeing, either within the eye or posterior to it, and then to apply stimuli of varying types and observe the effects on the cortex. But it must be kept in mind that the total path involves the retina with its bipolar cells, and ganglionic cells, the chiasma and optic tract, the thalamus, the optic radiation, and finally the cortex itself, with all of their inter-communications. Without doubt it is the inter-communications mentioned which result in nervous activity by other parts of the body which are secondary to the paths which lead to the cortex and which latter paths result in the act of vision.

Experiments in which the retina itself is used require that the head be fixed so that it cannot turn, volitionally, because such

3 Wright, ''Perception of Light,'' Chem. Publ. Co., 1939.

turning sets up action currents in the neck muscles which would interfere with interpretation of impulses in the brain. Also, the eyes themselves should not be permitted to turn in their orbits for two reasons: First, such turning of the eyes sets up action currents in the extra-ocular muscles which may set up confusing nervous impulses, and, second, turning the eyes causes the light stimulus to fall upon a different portion of the retina which would, of course, alter the type of impulse transmitted. Dilation or constriction of the pupil, activity of the ciliary muscle controlling focus, winking of the lids, and vasomotor changes in the retina itself are all brought about by nervous impulses to the respective parts, and unless they are carefully controlled will, to some extent, alter the type and kind of impulses transmitted to the brain cortex. It is true, however, that muscle sense impressions and even the state of contraction of the ciliary and pupil are a part of the nervous mechanism of vision, all of which enter into the mental interpretation resulting from the receipt of their impulses by the cortex. Obviously, if a stimulus is applied directly to the cut end of the optic nerve, just posterior to the eyeball, it is possible to arouse the pathways in the brain which result from this direct stimulation of the optic nerve and would to some exten overcome some of the difficulties mentioned above.

When light falls upon the rods and cones in the retina a chemical or a physical change takes place and the energy absorbed results in the elicitation of a nerve response which itself has previously been shown to be either purely chemical or of a physicochemical nature. Following this more or less local action the events which follow constitute the visual impulse, as directly contrasted with the stimulus of light energy. This visual impulse exists in one or more of the parallel fibers in the optic nerve and if a multiplicity of fibers are simultaneously stimulated we have a synchronized discharge as contrasted with the random discharge of fibers in the event of asynchronous stimulation of a multiplicity of fibers, regardless of whether the impulse in the fibers are synchronized or are asynchronous. Each impulse carried by the several fibers must "change cars," as it were, at each synapse along their several paths. It should be recalled that synapses occur at the bipolar cells, in the ganglionic layer of the retina, also in the thalamus and in the cortex. Also, there is evidence that a nerve impulse transversing one fiber may affect or institute impulses in parallel fibers. To completely understand the nature of and the factors involved in the sense of sight it will be necessary to know something of the impulses as they leave the rods and cones, where, admittedly, there is some chemical change. Of course, in the nerve fiber itself there is simply a nerve impulse, but at the level of the synapses it becomes complicated by the automatic cell impulses themselves, which are somewhat different from the nerve fiber impulses. Also, we should have to know something of the velocity of these impulses, their frequency and just what takes place in the visual act whether the impulses are the result of a stimulation of a single fiber, or if there are many fibers delivering impulses to the synaptic level and the frequency at which these impulses are delivered. Such knowledge would give a physiological description of the visual act, but would not describe the mental processes involved.

The optic nerve contains approximately eighty thousand fibers of varying cross sectional sizes. By a study of the relationship of the character of impulses and the ease of stimulation of nerve fibers of different sizes it is found that those of large diameter are easier to stimulate electrically and carry impulses at a higher velocity than fibers of smaller sizes. This fact lends some support to the previously mentioned "all or none" theory of retardation at the synapse. Furthermore, in practically all nerves that have been investigated it has been found that the actual nerve impulse is not a simple flow of energy, but occurs as a series of discreet impulses. It has further been found that these group sizes of nerve fibers vary in terms of the special function of their receptors or effectors, in that some fibers from the skin, for instance, conduct sensations of pain, whereas other fibers carrying only the sense of touch and the smallest fibers affect the autonomic and control the caliber of the blood vessels. In the cervical sympathetic trunks there is found one group of fibers which activates the pupils, another acts as a vasoconstrictor to the blood vessels, and yet a third group of fibers are purely sensory in function.

Bishop[4] reasons from the foregoing that since in all other nerves

[4] Bishop, "Functional Study of Nerve Elements in Optic Pathways," Amer. Jour. Opt., Nov. 1934.

we have a multiplicity of fibers with differing sizes and functional transmission by these fibers, that we should expect to find in the optic nerve a minimum of three functions, two of which correspond to potential waves of the first order following stimulation of the retina or nerve end, which have to do with the act of seeing as such, and a third, the *"motor visceral function control possibly affecting the general health."* Such a visceral function control would obviously indirectly determine the receptivity state of the retina by maintaining it in a proper physiological condition to transform light stimuli into nervous impulses or otherwise. *Bishop* continues by implying that the exact nature of these functions has not yet been fully investigated, nor have they been identified although there is sufficient evidence to indicate that these two sets of functions are distinct entities.

A cross section of the optic nerve shows fibers of a wide range cross sectional size. These fibers are not massed in groups, but are rather thoroughly distributed throughout the section. However, there is one region of the nerve which contains all large fibers. It is known that the spatial arrangement of the fibers in the nerve has a definite relationship to the ending of these fibers in the retina itself. And since there exists a difference of conduction rate in fibers of different sizes it is a logical inference that the two waves in the two size groups of fibers serve the sense of vision itself. This conclusion seems to be supported by the fact that electrical stimuli applied to the cut end of the optic nerve is distributed to definitely localized regions of the retina which correspond to specific locations in the nerve trunk.

An electrical stimulus applied to the proximal end of the cut optic nerve will excite a burst of potential responses by that nerve. This response is equivalent to and identical with the response of the nerve to a short flash of light which falls upon the entire retina and which is not focused on a point thereon. There is this difference in the experiment, however, a short flash of light may send a succession of impulses down the nerve along the fibers stimulated by the retina, whereas an electrical stimulus will only excite a single impulse. There is another difference in that an increase of intensity of the light will increase the number of impulses sent over each fiber, because the greater the stimulus to a sense organ the more rapidly it discharges impulses down the fibers leading from it.

When a single stimulus of low intensity is applied to the cut nerve it is usually not possible to detect a cortical response, probably because of high synaptic resistance enroute. However, there is another angle. If a larger number of fibers are stimulated the impulse goes through and is detectable in the cortex in similar manner to the effect produced by repeated weaker stimuli to a smaller number of fibers. If a single electrical stimulus has its intensity increased sufficiently, a point is reached where eventually there is a cortical response. In the former instance where a large number of fibers were stimulated with a weak stimulus we may have a response due to excitation of the larger number of fibers, the effect being comparable to increasing intensity of the stimulus over a smaller number of fibers. Since it is known that every electrical shock stimulates the nerve, but that some of these nerve impulses do not reach the cortex, the only assumption which will account for this fact is that somewhere below the cortical level, in all probability in the thalamus, a variation in sensitivity of response occurs which is independent of the stimulus itself. This modification probably takes place at the thalamic level as a result of delivery of impulses out of phase with the automatic discharge rate of the thalamus itself.

If a stimulus is increased until the larger number of fibers become activated a point is reached where every application of the stimulus results in response at the cortical level, but it is a notable fact that not all of these cortical responses have the same intensity. For instance, one impulse may have high intensity, and the two or three succeeding ones have a low intensity, also the fluctuations may be rhythmic or arrythmic. Since it is known that the optic nerve response does not vary in any such manner the only place where such variation could be caused is subcortically, probably the thalamus according to *Bishop* and *Hartley*.

If the strength of the stimulus is strong enough and the number of fibers stimulated is great enough the response at the cortical level is uniform in intensity and rhythm. In fact this is so true that every stimulus will excite a virtually identical cortical response. Any change in the intensity of the stimulus, or the rate at which it is applied, immediately results in the arhythmic fluctuations mentioned above. Obviously, these conditions could only result in the presence of a regular rhythmic nerve cell discharge *somewhere between* the point of application of the stimulus and the cortex. Such a natural periodicity of nerve cell discharge is known to exist. If an impulse reaches a cell in phase with a rhythmic rate, the result will be additive and will be conducted forward toward the cortex. *Contra* if the impulse is received at the cell out of phase with its automatic rhythm it may neutralize the rhythmic discharge completely or may greatly lessen the intensity of that discharge. If the rate of receipt of impulses is identical with the rate of rhythmic discharge the end result will always be additive and maximal, with all discharges identical in amplitude. This rhythmic stimulation frequency has the above mentioned additive effect when it reaches a frequency of approximately five per second. While electrical stimulation of the nerve end is not physiologically identical to the stimulation of the retina by light it is, however, found that light flashes in the eye, that are identical in time and in intensity, set up the same type of fluctuations as those discussed above, which are produced by electrically stimulating the cut nerve end.

All of the nerve cells in the cortex are constantly discharging a series of rhythmic impulses, but it must be kept in mind that they all do not discharge simultaneously. During the instant of discharge they are relatively refractory to the receipt of an impulse, but *contra*, they are apparently open to stimulation by an impulse which reaches them during the rest period between discharges. A stimulus arriving over an optic radiation tract to the cortex at any given point and time would probably only activate those cells which were at that instant in a resting state between their own spontaneous discharges. It follows, since all of the cells do not discharge simultaneously, that there will always be some in the resting state so that impulses arriving at the cortex will always find a receptive nerve cell. This would, of course, result in vision being a relatively constant phenomenon, there always being a receptive cell present to accept an incoming impulse. We thus see the possibility of some cortical cells resting or relieving others successively. A further bearing on this point is that stimulation of the optic nerve directly, or of the retina by light flashes of weak intensity, will cause a constant response *only* if they are delivered at the proper critical frequency.

Physiologically, we know that the large fibers respond to weaker stimuli and have a lower threshold than the smaller fibers. From the similarity of cortical response to an electrical stimulus applied to the cut nerve end, and to a light flash in the retina, the conclusion is reached that the larger fibers connect with more sensitive receptors in the retina, which correspond to weak light. These are commonly accepted to be the rods, and in accordance with *Houston's* findings, *supra*, may carry impulses from a multiplicity of rods, due to lateral conduction in the reticular layers of the retina. This conclusion is further supported by the finding that the larger fibers deliver impulses to cortical cells which act synchronously and in phase with each other, while the smaller fibers deliver impulses to cells which are out of phase with the former, but are necessarily in phase with each other.

Hartley makes use of the fact that the optic thalamus discharges rhythmically as does the cortex, even in the absence of optic stimulation, and says, "It seems probable that the activity of the cortex, spontaneous as well as that due to nerve stimulation, is largely controlled by the thalamus." Supporting this we do find that the character and frequency of the impulse reaching the thalamus is quite different from the character and frequency of the impulse reaching the cortex, therefore, something must have happened to the impulse in the thalamus, enroute to the cortex. It seems apparent that the summation of receipt over the multiplicity of fibers by the thalamus results in summation taking place therein, particularly when the impulses are too weak in themselves to influence the cortex.

Granit, quoted by *Bishop* and *Hartley*, has shown that the negative state of potential in the retina persists while the retina is exposed to light. He, therefore, concludes that the physicochemical process which causes the negative charge is the one responsible for stimulation of the optic nerve fibers. The change of the retinal charge to a negative one under exposure to light presupposes a standing positive charge on the retina in darkness.

Since a single discharge, and its amplitude, of an impulse passing down a nerve fiber always is a unit value, the only way it is possible to differentiate between different intensities of a stimulus acting upon a sense organ is by noting the frequency of the volley of impulses discharged down the nerve from the sense receptor. From this fact, and taken in conjunction with the foregoing above mentioned facts, we find that an *increase in exposure time* of the eye to a light stimulus acts in an identical manner, in so far as the cortex is concerned, as does an *increase of intensity*. Measurement of the receipt of impulses thus set up at the cortex is obviously not a measure of the mental or psychic response thereto, but is merely a measure of the physiological processes involved, the resultant of the stimulus and by the several modifying factors between the ocular receptor and the cortex. This modification takes the form of an altered frequency and, also, an altered amplitude as it passes each level of the pathway. We can see, therefore, that in the thalamus and in the cortex the impulse has met conditions not normally found in the periphery. This is because those nerve cells which must be passed enroute are already sending out continuous automatic waves of impulses, and are not wholly dependent upon the receipt of an external stimulus, such as light falling on the retina. The optic nerve, therefore, does not find quiescent nerve cells awaiting activation by its cell impulses, but must deliver them into a felt-work of cells that are already in a state of automatic activity. It follows, due to modulating effects in this felt-work, that the impulse created at the sense receptor must lose its individuality, and having reached the cortex, the interpretation of these visual impulses is one of interpretation in part of the impressed changes by the nerve cells traversed enroute. This, of course, reverses the older idea that the brain directs the periphery, but this reversal is not without support by the experimental evidence adduced.

McCullough,[5] in discussing the irreversibility of a single synapse, holds that a temporal summation at a single synapse is experimentally improbable, but that spatial summation by neighboring synapses is possible anatomically and has been demonstrated physiologically. His reason for this statement is that a subliminal stimulus does not set up a nerve impulse, therefore, spatial summation can only occur at synapses that are close together or in contact upon the cell receiving the impulses.

Consequently, if summation is required for cross neural conduction, and for temporal summation at any single synapse, it can not result, while spatial summation occurs from the main

[5] McCullough, ''Irreversibility of Conduction in the Reflex Arc,'' Science, 87, 2247, 65.

terminations of axones upon a single cell, but in the reverse direction.

Beams'[6] findings reveal that the length of the light flashes is not a vital factor in the way the eye receives light energy. He found that the eye sums up or integrates light energy, at least over the range of flashes he studied, which ranged from sixteen thousand to twenty thousand per second. It follows from the foregoing that Fletchner's law has only a limited application to the function of vision. It will be recalled that Fletchner's law states that "the absolute change in sensation is proportional to the percentage change in the stimulus."

In addition to the electrical changes found in the cortex and along the visual tracts, *Gerard*[7] showed that the temperature of the brain and its conducting paths is markedly increased by the stimulation of light to the eye, and that this increase in temperature takes place in the relatively short time of one minute, and that the temperature continued to increase for about two minutes thereafter, but that at the expiration of four minutes from the application of the light stimulus to the eye, the temperature of the brain had returned to its normal resting degree. This finding would indicate that the nerve cells increase their utilization of oxygen during activity states caused by light impinging upon the retina.

In connection with the effects of light upon the eye itself we find that wave-lengths shorter than 295 mμ do not reach the retina, and that they are highly irritating to the conjunctiva and cornea. Wave-lengths shorter than 1.4 μ are practically all absorbed by the cornea and aqueous, thus apparently ruling out infra-red as the cause of degenerative changes in the lens and retina which might impair vision. *Sheard* reported in 1924 that it is extremely doubtful if there is any specific action of ultra-violet upon the retina, although ultra-violet wave-lengths do cause fluorescence of some of the transparent ocular tissues and may thus by lessening saturation of the visual image impair vision during the time of exposure to ultra-violet.

[6] Beams, "Human Eye and Flashes," S.N.L., Nov. 3, 1936.
[7] Gerard, "Frequency of Brain upon the Reception of Optic Impulses," S.N.L., Sept. 5, 1936.

CHAPTER XIII

FUNCTIONAL CONTROL BY EYES

Any study of the living organism which in any manner considers its function must also give thought to conditions which may alter that function. In an earlier place it has been shown that function of the organism as a whole, as well as function of the cells and tissues of which it is composed, in part depends upon the maintenance of an electrical potential difference. It was there shown that a gradual loss of potential difference resulted in gradually decreased functional power and that when this potential difference approximated zero there was a complete absence of the ability to function or to live. Cellular and tissue function are a part of the general metabolic processes which by their very activity, either in a chemical or physicochemical manner, aid in the maintenance of potential differences. It is thus seen that the two processes in life are inseparable and to some extent inter-dependent.

Metabolism of the body as a whole, however, is fundamentally under the control of the autonomic nervous system, which in turn maintains a balance between the secretions of the endocrine glands in addition to the dynamically antagonistic control it exercises over organs composed of smooth muscle and certain other glandular tissues which are activated solely below the conscious level. During a state of dynamically equal control between the two divisions of the autonomic, a condition which we call *syntony*, we find that function within the individual is at its best. The use of the word syntony, or syntonics, in this connection was first made by *Blueler* in an article entitled "Die Probleme Der Schizoidie und der Syntonie," z. Neur., 78, 1922, in so far as the writer knows. Since that time the term has come to connote an integration of the nervous system for adequate and competent function.

Due to the fact that the two dynamically antagonistic divisions of the autonomic activate endocrine glands which are functionally antagonistic, it becomes apparent that only when a condition of nervous syntony exists will it be possible to have an adequate and normal endocrine status in the individual. There is one particular endocrine factor which should be mentioned here, and which shall be considered in detail later, as it pertains to the function of the pituitary, i.e., hypophysis. The role played by this gland may be one of purely nervous origin, or one purely hormonal, but, whatever the role, the effects will be shown to be very wide-spread, enough so almost to be said to be universal within any individual.

It has previously been shown that the central gray and the thalamus act as a nervous organization or distributing center for the maintenance of syntonic function between the two divisions of the autonomic. Also, that stimulation of thalamic function, under the stress of emotion such as anger, fear, rage, et cetera, as was shown in a previous chapter, and will be shown later in this chapter, can be caused by light stimulus itself, is found to activate the sympathetic division of the autonomic. Stimulation of the sympathetic automatically and instantly produces all of the effects listed in the sympathetic syndrome *vide supra*.

But there is another connection between the central gray and the endocrine system which is not purely nervous. This connection is the portal circulation arrangement existing between the central gray and the pituitary. By reason of this circulation, it is found that the central gray may directly, by nerve stimulation, affect the pituitary, while itself being chemically altered as to functioning power by the pituitary secretions being returned to it through its portal circulation. Obviously, such a condition could very quickly become what *Hurry* calls a vicious circle or a vicious cycle, because if such an interlocking process were to get very far from its normal manifestations, the two effects might become so fixed that they would become self-perpetuating regardless of external or internal environmental factors. Such a vicious circle would require more than physiologic measures for its correction and such corrective methods are obviously outside the scope of this work.

Since the act of living and general metabolism is a constant effort to maintain function by either balancing, or temporary loss of balance, between "driving" and "inhibitory" functions, it seems imperative that the relationship between nervous and chemical drives be here discussed. Any stimulus which disturbs the relationship between the sympathetic and the parasympathetic, or which causes an excess of secretion by any one or more of the endocrine glands, is capable of temporarily causing malfunction. Temporary malfunction continued over a period of time results in a class of conditioned responses which become more or less permanent and must be retrained if the originally normal relationship is to be brought about. Among the men who have made special study *Breitmann* of Leningrad, has devised a clever schema to show the relationship of the synergism and antagonism between the several divisions of the endocrine system, the nervous system, and the other glands and tissues of the body. This schema takes the form of three broken concentric

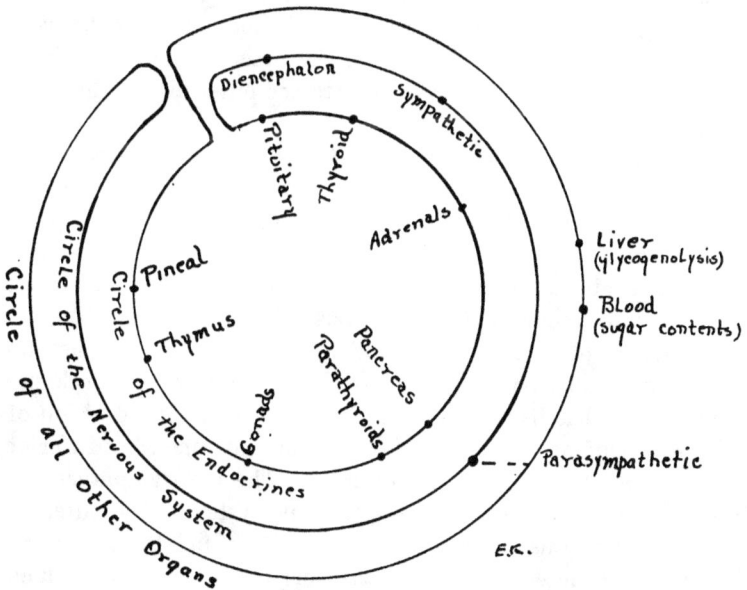

MODIFIED FROM BREITMAN
Figure X.

circles. In the inner circle, beginning at the top and moving clockwise therefrom are stationed the endocrine glands in the order of their dominance. Those placed higher in the circle have a stimulating effect upon the one next below, and *contra*, those lying lower have a contra-stimulating or inhibiting effect upon those placed higher in the circle. It should be noted here,

however, that there are no known primarily inhibiting functions
exercised by the pituitary. Any such effects attributable to the
pituitary must be secondary as a result of its stimulation of some
other gland which antagonizes the gland apparently inhibited.

At the top of this inner circle of *Breitmann*'s we find the
pituitary—hypophysis. Next, moving clockwise, the thyroids,
adrenals, pancreas, parathyroid, gonads, thymus and epiphysis
in that order. The last two have somewhat analogous effects,
and the present state of knowledge does not permit giving one
preference over the other so they are listed very closely together.
There is no proof of directly reversible relationship between any
of these glands, hence the inner circle is not completed between
the epiphysis and the hypophysis, the break being respectively
connected to the second circle, and to the outer circle, the third
one.

The second or middle circle of *Breitmann*'s schema contains,
at the top, at the highest level, the diencephalon or the central
gray, that portion of the brain which dominates the autonomic;
and moving clockwise we find the sympathetic system next in
order, probably due to its control of those functions which are
definitely necessary for instantaneous integration of the body in
the presence of danger. This action, as has been shown heretofore, is a mass pattern action. Proceeding further in a clockwise
direction, he has listed the vagus or parasympathetic division of
the autonomic at a somewhat lower level, but somewhat inferior
to the sympathetic at its higher level. The parasympathetic is
then followed by the rest of the nervous system. The outer or
third circle of this schema contains the liver and the co-related
blood sugar content, and the other organs of the body which
enter into the maintenance of the complete life cycle.

Attention should here be directed to the fact that the two
highest levels are the pituitary or hypophysis, the controlling
gland of the endocrine system always being primarily an activating gland; and the diencephalon in the brain, which is the
integrator of the autonomic system. The hypophysis and the
diencephalon interlock in two ways as has been stated above:
First, direct nervous control; and second, a portal circulation
connecting the two. By this dual connection, influence of either
one calls forth definite response in the other, and due to the

multiple ramifications of the autonomic under the control of the diencephalon and the endocrine glands, largely under control of the hypophysis, a wide-spread influence is maintained over all function in health as well as in disturbed function during disease.

As has been seen in an earlier chapter the pituitary, particularly the anterior portion, controls the sex cycle in animals. The sex cycle in the female is observable by checking on the rate of ovulation, which in human female has a normal 28-day intermenstrual cycle. It has been found in animals that destruction or removal of the pituitary abolishes sex cycles and ovulation, seasonal shedding of hair, and other somewhat more obscure effects. A peculiar association with the pituitary lies in the fact that cutting of the optic nerves of animals in which the pituitary has *not* been destroyed, results in abolishing the hair shedding cycle and alteration of the sex cycle from a purely seasonal occurrence. This indicates some tie-up between ocular absorption of radiant energy and pituitary function. It has long been known that animals depend upon daylight, or perhaps more properly speaking, day length, with increased intensity of light for the inauguration of seasonal sex stimulation. In the northern hemisphere this cycle becomes evident in the spring of the year as is easily observed in domestic animals such as the horse, cow, pig, chicken, turkey, and in wild life, such as the rabbit, hedge-hog, frog, fish and bird. As mentioned above the length of the day, and the intensity of the light, are contributing factors to these phenomena, but it should also be noted that these effects are to a large extent dependent upon the wave-length, frequency, of the light.

Rowan[1] during his study of migration of birds, in reference to light intensity, found that as the duration of the day, and the intensity of illumination increased, birds tended to migrate northward into the more northern parts of the hemisphere. His most interesting finding was that by exposing birds to artificial electric light, he was able to change the time of year in which the birds would undertake migration, thus proving that it was the duration of the light, possibly primarily, and an altered frequency, probably secondarily, which are controlling factors in the migration drive impulse. By the same means *Rowan* also

[1] Rowan, Nature, 119, 351.

altered the time of year of egg-laying, and hatching by the hens. At first it was thought that the temperature of the surroundings might have something to do with migration and egg-laying effort. In an effort to check up on this possibility, *Rowan* used his controlled light sources and temperature as low as – 50 degrees F. and in even such low temperatures was able to prevent the egg-laying and hatching periods, or the attempted migration, when he increased the duration and intensity of his light sources. He thus proved that it was the light itself which was the major factor in the phenomena mentioned. *Bissonnette*[2] was able to alter and to control the rate of the sex activity in starlings in any time of the year by merely exposing them to light for longer or shorter periods of time. The rate of sexual activity varied directly with the intensity of the light. He found that the *frequency* of the light was also a very prominent factor and that the altered sex activity was greatest under red light, while there was no alteration under green, and in his experience probably none at all under blue. Other researches indicate that blue actually has a retarding effect, which researches will be cited in the proper place herein. *Daken*[3] and *Bissonnette* are both in agreement that a proper food supply is necessary in all instances in which light is used for altering the functions mentioned above.

Energy content measurements show that the wave-length of light itself has its specific response, which is *not* a function of its total energy content. In other words, it is the frequency of absorption of quanta by the receptor which determines the result and not in the amount of energy absorbed as a total. It was also found that a cataract on one eye of an animal or bird did not materially alter the physiologic responses to light and it is a well known fact that human beings with cataracts usually retain their ability to distinguish the extremes of the spectrum. Consequently, as a result of the above discovery by *Bissonnette*, he suggested on February 1, 1932, that the light stimulus acted *through* the eyes and the central gray, thereby affecting the anterior pituitary and through it the adrenal cortex and the gonads. *Hill* and *Parks*[4] confirmed the above conclusion by

[2] Bissonnette, Proc. Assoc. Res. Nov. Dis., XXV, 316.

[3] Daken, personal connection.

[4] Hill and Parks, Proc. Roy. Soc., 113, 530–544.

Bissonnette by removing the pituitary gland, and discovered that all of the cycles above mentioned were no longer produced nor altered by the changes of light in the eye. This definitely lodged the responsibility for control of these functions with the pituitary gland, acting as the mediator of energy or impulses from the eye as a result of light stimulus. A most interesting finding by these men was that even if a very minute piece of the pituitary were permitted to remain during the ablation operation, the animal so mutilated still responded to the effects of light in the eye, but, that, where ablation of the pituitary was complete, there was no response as cited above.

Marshall and Bowden[5] confirmed the findings by *Hill and Parks* and further found that a *weak light stimulus over a longer period of time was equivalent in effect to a strong light stimulus over a shorter period of time.* They reduced this finding to the formula that if the product of *intensity* times *time* are the same in both instances, the physiologic responses would be identical, $IT = I'T'$.

Bissonnette and *Wardland*[6] reported that in a ferret with double cataract there was an absence of sex cycles for a period of three years. In an effort to associate this with ocular stimuli, they cut the optic nerves behind a pair of normal eyes in a ferret, following which operation, oestrus was missed for about two years, but that in the third year there was an oestrus, indicating that there exists an inherent within the animal mechanism, in addition to light perception and stimulation, which operates to maintain sexual activity of the animal under the conditions named. They found that normal ferrets responded to light in the infra red region, through the visible range, and into the ultra-violet. It should be noted here that ferrets are nocturnal predatory animals, and are obviously much more sensitive to low degrees of illumination than are starlings, which *Bissonnette* found to respond more to red light. Destruction of the pituitary permanently destroys the normal oestrus in the ferret as well as the normal shedding of hair. Cutting of the optic nerves *also* destroys shedding under the influence of light, and, as has been mentioned above, destroys oestrus experimentally for a period of over two years. *Benoit*[7]

[5] Marshall and Bowden, Jr. Exp. Biol., 11, 409, 422.

[6] Bissonnette and Wardland, Jour. Morph., 52, 403–428.

[7] Benoit, Acad. Sci. Paris, 195, 1671–1673.

found that red was effective in increasing ovulation in ducks, and that green and blue were not effective, thus confirming *Bissonnette's* findings *vide supra*. *Benoit* also found that removal of the pituitary destroyed the effect of red.

Researchers now wondered if the above mentioned effects were produced by light falling upon the skin areas, or if it were due solely to the effect of light in the eye. *Bissonnette* made masks with small holes for the eyes covering the whole heads of his animals and birds. He thus excluded from exposure to light all parts of the animal except the eyes. The result of this experiment proved conclusively that the eyes were the *sole* source of nervous impulses and were the only receptors of the light stimulus which resulted in the responses previously noted. *Bissonnette* also tried the same experiments on ferrets and ducks, starlings and monkeys. The writer has confirmed *Bisonnette* in this connection with reference to monkeys, i.e., that it was the eyes alone which were the sole receptors of the light energy which resulted in the physiologic changes.

Ivanov[8] reported that light played upon the freshly cut end of the optic nerve in the orbit was partially effective in producing physiologic responses, but that if the orbit was plugged with a rubber dam and light then played into the orbit, but not permitted to reach the cut end of the nerve, all responses were stopped. He thus proved that the effects on the pituitary were by the way of the eyes to the optic nerves and optic tracts to the central gray.

Collins[9] described nerve fibers going to the pituitary from the supra-optic nucleus, the central gray substance and the tuber cinereum. He regards the oculopituitary reflexes following light stimulus to be largely a function of the thalamus itself. Incidently, he found that there is definite evidence of histological changes in the pituitary following the nervous stimulation, by modified exposure to light, and that these changes are in the basophil and eosinophil cells in the anterior pituitary which show vacuoles such as are found in castrated animals. *Jones*[10] found that the melanophore changes are abolished if either of the eyes

8 Ivanov, Arch. of Exp. Pathol. and Par.

9 Collins, Soc. Biol. Dana, 118, 1560.

10 Jones, Klin. high Sch., 14, 1713, 1716.

or the pituitary are removed. He found in the rabbit, kept in darkness, that the melanophore stimulation hormone is increased in the blood and reduced in the pituitary. In light this hormone is stored in the pituitary. He found that in frogs, blinded by enucleation of the eyes, but having the cut end of the nerves stimulated electrically or directly by light upon the cut end of the nerves, reacted like normal frogs in reference to melanophore changes.

Rudenwald[11] found that some of the several fractions of the pituitary secretions were altered, particularly the vasopressor and the uterine contracting fractions, were doubled and in some instances trebled in darkness or in blindfolded animals. The melanophore stimulating hormone was found to decrease in darkness whereas the vasopressor and the smooth muscle contracting hormones increased in darkness. *Rudenwald* reasoned that the vasopressor is in inactive state of the melanophore hormone, which might account for the fact that more births take place at night than in daytime. The writer wishes to add the following observation, since more deaths take place between midnight and morning that the increase in the vasopressor fraction, with its lowering of circulatory tonus, may be directly responsible, due to the lack of proper and adequate light stimulation during this part of the day.

The only logical conclusions in conformity with the facts which can be drawn from the foregoing is that the amount of light, and its wave-length received by the eye, and mediated by the optic nerves and thalamus, affect the pituitary, and is accompanied by physiological changes in that gland and in the degree of its activity. There are also, well recognized psychic changes in the individual as a result of light stimulation of the eye. The pituitary is commonly recognized as the master endocrine gland, as some one has said, "The chairman of the endocrine board of directors." Anything that affects the "chairman" has a far-reaching and highly involved series of effects upon the animal, which involve its structure, the length of its bones, its hair distribution, its sexual parts and reproductive cycles, its metabolism mediated by the thyroid, its energy release mediated by the

11 Rudenwald, Z. rerg. Physiol., 21, 767.

adrenals, and in fact its entire physiological and ecological response.

Daken[12] gives some quantitative findings as a result of his exposure of pullets in the same age groups to light of different frequencies. He finds that egg-laying increased 10.6 percentum in pullets which were kept constantly under red light as compared with those that were constantly under white light. Contrasted with this finding under red light, he found 20 percentum decrease in egg-laying in the group of pullets kept constantly under blue light, and practically no change in egg-laying by the group under yellow light when compared with those under white light. The light sources used in all of the four groups—white, red, yellow, blue—were all artificial, and the pullets were kept in a place screened from daylight, artificially ventilated and air conditioned during the experiment, which extended throughout a period from the first of August of one year to and including the middle of January of the following year. *Daken's* findings, due to the fact that his experiments were well controlled and all other light factors were excluded except those in use, is a beautiful confirmation of the findings of *Bissonnette* and *Benoit*.

Daken reports two most interesting side-effects in the pullets in the pens exposed to *red* light. One was that they developed a very high degree of cannibalism, in fact to such an extent that 84 percentum of all the pullets were more or less injured or damaged during the experiment. The other side-effect was that all of the pullets refused to roost on the roosts provided, preferring to sit on the floor or in the nests as far removed as possible from the sources of red light. Later they were mechanically restrained and forced to sit on a roost under the red light. Birds so mechanically restrained developed bronchitis with a high mortality before the conclusion of the experiment. These two side-effects apparently did not alter the ovulation rate to any great extent, or, if it did, the figures still indicate that there were more eggs laid by these pullets despite the side-effects mentioned.

It is quite generally recognized that of the two divisions of the autonomic the sympathetic division activates the posterior lobe of the pituitary and most authorities are in agreement that it also activates the anterior lobe of the pituitary, thus showing a rather

---

[10] Daken, Poul. Trib 11, 34, et seq.

complete domination of this gland by the sympathetic aside from the chemical control of the gland due to other endocrines in solution in the blood stream.

In the foregoing it has been shown that the incidence of light, or of certain light frequencies in the eye, exercises a very definite effect and control over the hypophysis which, due to its interlocking nervous and endocrine gland ramifications, causes widespread organic and nervous function effects necessarily to follow. A study of these from the standpoint of the system as a whole, and the general health of the individual, as well as the effect upon the function of vision and the control thereof, will be considered later.

CHAPTER XIV

BODILY HEALTH

Throughout Parts I and II, numerous mentions have been made of the effects of light on living tissues. Some of these references have been in terms of mere cells, others in terms of tissues, some in terms of response by the skin to irradiation, and quite a few have been in regard to ocular and physical responses to light in the eye.

These references have discussed the mechanism of responses in the higher organisms by citing nervous reactions and correlated endocrine responses, but few details have been given as they pertain to the more specialized responses necessary to the maintenance of health. In this chapter effort will be made to go further into those mechanisms that have been found clinically and experimentally favorably to affect departures from the normal, which have not gone far enough to become structural changes.

Professor A. M. Low,[1] a British authority, even went so far a few years ago as to recommend replacing the bodily energy, lost by the act of living, by direct irradiation of the skin by light from some electric source. His contention was that the energy consumed in physical effort which was not immediately replaced by the processes of metabolism might well be absorbed by the skin from radiant energy. "Take a light bath instead of a meal" might be a semi-serious summation of *Dr. Low*'s suggestion. Unquestionably energy is absorbed by the skin, but to the mind of this writer it is open to debate whether radiant energy absorbed by the skin could replace the reserves used by the activities of life.

That the absorption of light by the covering of the body can prove to be dangerous to human beings is proved by the news reports in the papers and in the medical literature of damage, even death, following prolonged exposure to the sun. In the case of human beings this exposure results in the formation of certain chemical compounds in the skin which, when taken into the circulation, may damage the kidneys. Blonds and red

[1] Low, Pop. Sc., Mar. 1934.

headed persons are particularly susceptible to this form of damage. *Mosauer*[2] reports his finding in respect to snakes and lizards which are normally inhabitants of the desert countries of United States. He found that they were killed by sunlight in from twelve to twenty minutes when lying on the ground. The first thought was that it was the heat of the ground due to absorption of the sun rays which was responsible for the death. The experiment was repeated with the reptiles supported about five feet off the ground and much to the surprise of *Mosauer* and his associates the straight insolation killed the reptiles in about the same length of time as was the case when they were resting on the ground. The conclusion reached was that absorption of the sun's rays was the *modus operandi* leading to death rather than the heated surroundings. *Dr. Frederic A. Woll* of Columbia University and City College, New York, verbally confirmed to the writer that his findings had been the same as those of *Mosauer,* using the side-winder rattlesnake as the subject of the experiment.

Rowan[3] reported that if starlings were forced to physical activity during irradiation by light there resulted an accelerated breeding cycle which he attributed to stimulation of the pituitary gland and the sex gland. *Rowan's* findings supplement those cited in Part II. *Pigett*[4] reports that he has demonstrated a connection between the pituitary gland and epileptic convulsions. In epilepsy there is too much water stored in the body in similar fashion that too much sugar is found in diabetes. The pituitary is known to produce a hormone which will cause the presence of too much sugar in the blood. *Contra,* under-function of the pituitary gland in reference to this sugar producing secretion would cause too little blood sugar, technically known as a hypoglycemia, resulting in convulsions. *Pigett* recalled that *Cushing,* of Yale, was able to prevent convulsions in epileptics by the therapeutic administration of pituitary extracts which increased blood sugar, but apparently had no effect upon the stored water. *Pigett* reasoned from his findings and those of *Cushing* that epileptic seizures might be controlled by stimu-

[2] Mosauer, Science, June 16, 1936.

[3] Rowan, S.N.L., Aug. 21, 1937.

[4] Pigett, S.N.L., May 25, 1935.

lating the sugar forming hormone in the pituitary or by artificially increasing the blood sugar. Following this conclusion
by *Pigett,* one may well imagine his surprise and pleasure when
he discovered an epileptic whose convulsion frequency was reduced three hundred percentum after the patient became dia-
betic. During an epileptic seizure practically all of the skeletal
muscles enter into clonic spasm which is quickly followed by
tonic spasm. These convulsions are comparable in finding to
those produced artificially by insulin shock, a recently developed
treatment used for the cure of schizophrenia.

Gault[5] reported an interesting observation to the effect that
increasing the light in the room actually increased hearing
acuity. He found that an increase of light corresponding to the
use of four hundred watts more electric current was sufficient
to make a marked difference in hearing acuity, but in the cita-
tion at hand *Gault* gave no quantitative figures as to the increase.

Most of the foregoing citations apply to an irradiation by
"whole white light." *Perdrau,*[6] reported by *Behari,* found that
methylene blue killed certain disease viruses, *only* in the presence of light, but had no effect in darkness. Methylene blue dye
absorbs the low frequency end of the spectrum and the conclusion reached was that the high frequency transmitted by methylene blue proved to be the destructive agency. Incidentally, the
reader should be reminded that one might as well say that the
absence of the low frequency energy resulted in their destruction, as has been discussed hereinbefore.

Bodily health depends upon syntony of all functions. By this
is meant that the functions are quickly and evenly balanced, so
the degree of stability and excitability of nerve cells and other
tissues is maintained, the latter being largely the determining
factor in the maintenance of health. Since the nervous system
is the correlating agency in the body, and since it is the agency
which maintains and operates functions and keeps them within
physiologic bounds, it is apparent that the degree of stability
and the degree of irritability of the nervous system, particularly
the autonomic division, are the determining factors in the maintenance of health or the cure of disease. Please bear in mind

5 Gault, S.N.L., Sept. 16, 1933.
6 Perdrau, reported by Behari.

that it is utterly impossible to think physiologically in regard to the autonomic nervous system without considering the co-acting endocrine glands, particularly the pituitary, adrenals, thyroid and gonads. Special stress should here be laid upon the effect on the pituitary since it is considered to be the master gland of the endocrine system, and by its activity or lack of activity determines the quality of the secretions of the other endocrines, as well as exercises a wide spread control over metabolism.

It should now be clear that lack of syntony between the divisions of the autonomic or in the endocrine system will result in functional departures from the normal. These might not, because of their reversibility, be undesirable were it not for the fact that continued abnormal function may and often does result in structural changes which are not reversible, *Brown*.[7]

In Part II, during the discussion of the biotypes, it was shown that hereditary dominance of one division of the autonomic over the other would result in the development of rather definite structural biotypes. It was also shown that by a proper application of the criteria for the determination of the biotypes it is possible to learn which division of the autonomic became dominant or over-active in early life, either due to hereditary tendencies or as a result of some environmental factor. The contention of the present writer is that since the pituitary largely governs the rate of growth and the total growth of bones, and since the pituitary has been shown to be amenable to control by the use of some light frequency, that *one* of the environmental factors responsible for altering the biotype away from the hereditary tendency is undoubtedly the light environment in early life. Anthropologically speaking, any biotype, other than that classified as the syntonic heretofore, is a departure from the normal and such an individual is not physiologically in equilibrium, i.e., syntony. The proper procedure, therefore, resolves itself into the establishment of equilibrium of function *within the limitations* of structural biotype, the continued maintenance of this equilibrium being the maximum desideratum for the continued health of the individual.

It follows from the foregoing that controlled light frequency

[7] Brown, ''Sympathetic Nervous System in Diseases,'' Oxford, Med. Pub., 1923.

might well prove a preventive of ill health, particularly such disease processes as may result from faulty autonomic or endocrine function.

The localization of a disease process within the body of a specified person is largely determined by factors within that body. One of these factors is the specialized nerve supply received by the various organs. This specialized nerve supply is known to be adversely affected by environmental or traumatic accidents. Nervous shock or over-stimulation may, due to synaptic memory, become a potent factor in the determination of localized disease processes. The common neuroses are an almost perfect example of the damage which may be done by synaptic memories, in that any or all of the symptoms of disease processes may be produced by certain environmental stresses and shocks.

Since the light stimuli received by any individual are never the same as those for any other individual, it becomes possible to visualize numerous departures from the normal resulting therefrom, obviously, these departures would not assume the same form in any two individuals, due probably to the three interlocking factors, the type and kind of light stimulus, the biotype, and lastly, the heredity. It seems clear that mass causes of ill health might be chargeable directly to the frequency of radiant energy entering the eye, operating through the mechanisms heretofore discussed.

Following will be found a number of recognized diseases or functional conditions resulting from stimulation of, dominancy of, or over-activity of the parasympathetic:

Visceral hypermotility
Spastic constipation
Intestinal stasis
Mushy stools
Incontinence of urine or feces
Excessive hunger
Chordee
Bradycardia, excessively slow pulse
Hyperchlorhydria
Eye strain accompanied by nausea and headache
Hay fever

Asthma
Excessive bodily sweating
Increased contractions of muscles in the body of the uterus
Hypotension
Hypothyroid
Spasmodic laryngitis
Croup
Eczema
Enuresis
Migraine
Glaucoma—*Rivers*
Cystitis
Diabetes
Rheumatoid arthritis
Urticaria

The following conditions are accepted as being due to over-activity of, stimulation of, or dominancy of the sympathetic:

Acute diseases with fast pulse, rapid respiration, increased temperatures
Hypertension before arterio-sclerosis
Hyperthyroid
Chronically dilated pupils
Uterine cramps, due to os contraction
Retinal hemorrhage
Osteocarthritis
Hypertensive myocarditis
Catarrhal gastritis
Angina pectoris
Gastric ulcer
Atonic constipation
Dysuria
Dysmenorrhea
Tachycardia

Many conditions listed in the foregoing tables are recognized as symptoms and not as separate diseases. It is a well recognized fact in therapeutics that control of the presenting symptoms more often than not leads to control of the disease process. It is obvious that adequate therapeutics require an accurate diagnosis and a knowledge of the underlying cause of the symptoms, but it is not our purpose to discuss diagnosis or treatment, the sole purpose being to submit evidence in support of the writer's contention that altered light environment in the visible range is *one* of the factors leading to ill health. *Contra* a proper control of the light factor should lead to restoration of health.

Summarizing the foregoing light effects, they may be stated in the two following basic statements:

During departures from the normal

1. Low frequencies stimulate the pituitary, decrease the leak in potential, and tend to stimulate the sympathetic, producing physiologic activity of the defensive type.

2. High frequencies depress the pituitary, increase the leak in potential, and tend to stimulate the parasympathetic, producing physiologic rest or the vital type activity.

CHAPTER XV

ULTIMATE CONTROL OF OCULAR FUNCTIONS

It is commonly accepted that the function of accommodation is inherent by reason of the fact that it is governed by the parasympathetic division of the autonomic, and *contra*, obviously, is inhibited by over-activity of the sympathetic, a mechanism for the latter having been heretofore cited. There seems to be some question as to the ultimate mechanism of the accommodative function within the eye itself, due to the absence of fibers in the ciliary body which are recognized to be antagonistic fibers to those which are in action during the act of focusing at the near point. Some authorities assume that the elasticity of the lens substance is a mechanical opponent to ciliary contraction under the influence of the parasympathetic. Should future research prove this to be the case, the writer offers as an explanation for the inhibition of accommodation the probability that inhibition of the parasympathetic supply to the ciliary body is by a direct sympathetic inhibition at the ciliary ganglion, which is known to receive sympathetic fibers. The writer admits that this theory may not fit some of the known facts of loss of accommodative ability in old age, because the sympathetic is known to be less active in old age than in youth. Yet he sees no direct antagonism between this theory and the lens elasticity theory of loss of accommodation.

It is a well known fact that sympathetic activity resulting from a temporary paralysis of the parasympathetic by belladonna, or by some of its derivatives, results in a temporary loss in the power of accommodation. Adrenalin being a direct stimulus of the sympathetic, also lessons the power of accommodation, but in the latter instance the effect is much more transitory.

In view of the foregoing one would expect a more active accommodative function in a pyknic biotype, in which the parasympathetic is over-active. Clinical observation shows this often to be the case. Not only that, but clinically more pyknics are found to be myopes than asthenics.

Beasley[1] has conducted a series of experiments in an effort to determine the age at which an infant can be proved to have a

[1] Beasley, ''Baby Can See,'' S.N.L., Jan. 13, 1934.

sense of vision. He found that infants at early ages do see, as evidenced by the fact that their eyes exhibit the versions by following a moving light, thus proving that they see the light, even if it is not focused sharply on the retina by the act of accommodation. *Beasley* also found that the infant's eyes fixated moving objects such as fingers or a dark blue moving object, but he found no way of determining if infants could see stationary objects. Retinoscopic examination of infants' eyes indicates them to be highly hyperopic at birth, which condition would require a high degree of accommodative ability early in life if the error is to be compensated for and if clear vision is to result. Clinically this is found to be the case.

Bernstein,[2] later confirmed by *Steinhaus,* finds that the amplitude of accommodation is a rough measure of life expectancy and is somewhat directly proportional to the expectancy. Thus a low amplitude in early life would tend to indicate a relatively short life. Since the amplitude of accommodation indicates either inability of the parasympathetic to contract the ciliary body, or insufficient vital force to maintain a contraction in the presence of nerve stimulus, it appears that there is good physiologic ground for *Bernstein*'s findings, since the parasympathetic is the division of the autonomic which maintains and operates those functions vital to life.

The function of convergence is commonly accepted to be under the control of the cerebrospinal nervous system and that an accurate convergence fixation is a learned or acquired response. Some authorities hold that, in view of this, the act of convergence is merely a conditioned response due to long practice and training. If this proves to be true, by further investigation of the convergence function, it should be found that stimulation of the sympathetic would directly impair the ability of accurate fixation, and would show a lessened in-turning power of the eyes and a decreased ability to cause the visual axes to meet at the point of fixation. Clinically, this is exactly what is found following stimulation of the sympathetic by any of the common drugs which are known to cause it to become active, i.e., more active than its antagonist. It has been found that the accurate response of this

2 Bernstein, ''Lens Accommodation and Life Expectancy,'' Science 83, 2140, 9.

function is also impaired by emotional stresses and strains which are known to activate the sympathetic. Furthermore, the well known "sentinel response" resulting from sudden noises, which might indicate the approach of an enemy, also impairs the convergence function. Obviously, the approach of an enemy, indicated by a sound, causes the sympathetic to become dominant for the purpose of safety in flight, or as a defensive response. It is not denied in any of the foregoing that there is not some association betwen the function of accommodation, automatically controlled, and the function of convergence, cerebrospinally controlled, yet it is proper to state here that there is no *fixed* association between the two functions, further, one can be caused to be more active than the other by suitable optical or chemical means.

The size of the pupil is controlled by the autonomic, it being dilated by the sympathetic and contracted by the parasympathetic. A large pupil on one side is almost invariably a sign of misbehavior of the body, somewhere below the level of the brain. Of course, it is possible that sympathetic irritation may exist along the path of the sympathetic fibers within the brain, but, clinically, this is not as common a finding as misbehavior elsewhere in the body. *Salmon*[3] says bilateral pupillary dilation accompanies generalized stimulation of the sympathetic, and *contra* bilateral contraction of the pupil indicates general inhibition of the sympathetic. In the presence of any difference in the size of the pupils, or pupils unduly dilated or contracted the indication is that the case is not a purely refractive one and that search elsewhere should be made for the cause of the difficulty. *Aidie* has the following to say in "The Practitioner":

The expressions "pupils peculiar to syphilis" and "Argyll Robertson pupil" are not synonymous in common parlance, as they would be if strict attention had been paid to the original description. The pupils peculiar to syphilis have the following characteristics: 1. The signs are present in both eyes; 2. the pupils are small; 3. the light reflex, direct and consensual, is absent—the pupils do not contract with increased illumination or dilate when they are shaded; 4. the contraction of the pupils on covergence-accommodation is

3 Salmon, "Pupillary Prominence," Am. Jour. Obs. and Gyn., 28, 3.

prompt and complete; when the act of convergence ceases, the pupils regain their resting size at once; 5. the response to a mydriatic is slow and incomplete; 6. useful vision is present. If these criteria are all satisfied, the patient under examination is a syphilitic.

It has long been taught that the pupils contract when the function of accommodation becomes active. There is serious doubt that this is a statement of fact as indicated by the work of *Eidelberg* and *Kestenbaum*,[4] who have adduced experimental evidence in regard to fourteen phases of the act of convergence and its effect on pupillary contraction. They, therefore concluded that the pupils only contract in association with convergence and do *not* contract in association with accommodation. In other words, it seems that the cerebrospinal function of convergence may and does increase parasympathetic delivery to the pupil, or inhibits sympathetic delivery to the pupil.

Since man is endowed with two eyes, it is essential that a study be made of their mutual inter-activity. First, for consideration there is the ability to attain single simultaneous binocular vision with two separate eyes functioning as individuals. The attainment of this end requires that the visual axes meet at the point of fixation, or *that there be some cerebral mechanism whereby two images are fused as one if the axes do not meet at the fixation point*. It is easier to think, of course, in the former manner, and, incidentally, it is easier clinically to develop techniques based upon the meeting of the visual axes theory, nevertheless, it must be admitted that the latter alternative above may be, and undoubtedly is a very prominent factor. A second resultant due to the placement of the two eyes is that of the so-called "third dimension effect," i.e., stereoscopic vision. *Washburn*[5] reported to the National Academy of Science that the third dimension effect was primarily a motor experience and that it was not by a static fusion of retinal images, but resulted from *motion of the eyeballs themselves*. *Washburn's* findings have been recently confirmed by research work conducted at the University of Michigan by *Heinz Werner* wherein it was found that stereoscopic vision only takes place when the visual axes are moved about over a station-

4 Eidelberg and Kestenbaum, Jour. f. psychiat. u. Neurol., 46, 1, 1928.

5 Washburn, "Three Dimension Effect," S.N.L., May 6, 1933.

ary field and that it does not take place effectively when the field is in motion, with the eyes causing the visual axes to move about in efforts to follow the moving field. *Moriarty*[6] states definitely that the three dimension effect is not determined by the eyes, but that it is a function of the brain and cites an experiment in his article in proof of his statement. In addition to the above, the excerpt writer in ''Produce Engineering,'' finds that the color of the test object alone may cause as much as ten percent. variation in the estimation of distance, .distance estimation a recognized stereoscopic function.

Nutritional states of the eyes are known to alter not only visual function, but the functions of the extraocular muscles. The sympathetic is known to cause vasoconstriction of the blood vessels in the head. In the presence of vasoconstriction there is obviously a lessened supply of food and oxygen to muscles which would result in ineffective function. The superior cervical ganglion of the sympathetic is known to control this vasoconstriction in the head, but the superior cervical ganglion is controlled by the white rami communicantes leaving the spinal cord at the first, second, third and fourth thoracic intervertebral foramina. It seems, therefore, that there is some justification from the control of ocular functions by manual procedures in this region. Furthermore, head position in the lateral plane of the body may be such as to produce stretching with contraction of the carotid arteries, thus mechanically lessening blood supply to the head with consequent altered function.

It seems specious to mention the effects of endocrine glands upon ocular muscles, yet it must be mentioned that the ability of the striped muscle to go into rapid action and its ability to recover from fatigue is largely a result of adrenal activity. Furthermore, the ability of smooth muscle to hold contraction is largely due to blood chemistry and the presence of an adequate quantity of the secretion of the posterior lobe of the pituitary. The adrenal secretion is mediated by the sympathetic, whereas the secretion of the posterior lobe of the pituitary is apparently mediated by the parasympathetic. In this respect a secretion generated by the parasympathetic acts synergistically with the

6 Moriarty, ''Eyes and Third Dimension,'' S.N.L., Aug. 6, 1938.

parasympathetic division of the autonomic nervous system itself to maintain accommodation in accurate focus.

Haldane[7] reported to the Edinburgh Royal Medical Society that the perception of color was as much a function of the eye as of the particular frequency of the light incident thereto. This finding does not appear to support some of the former theories of color vision unless one admits that ocular structure is the thing that governs color perception, which admission seems to be of very doubtful propriety. Recent work by *Erickson*[8] indicates that color blindness may not depend upon either the eyes or to a great extent upon the frequency of the light. *Erickson* finds that color vision may be purely psychic as a result of his experience with subjects under hypnosis, during which it was suggested that upon awakening the subject would have no perception of certain selected colors. Subsequent re-hypnosis restored this perception.

There are many theories relative to the causes of cataract. Some hold that senile cataracts are caused by parathyroid, thyroid, or gonadal altered function. Others hold that it is merely a deposition of calcium salts in the crystalline lens of the eye. It is not the purpose of this work to evaluate any of these theories because the fact remains that a high percentage of cataracts can be favorably affected by the use of a selected frequency band in the blue-green range of the visible spectrum. *Mitchell*[9] found that the addition of lactose or beta-lactose to the diet invariably produced cataracts in rabbits. It appears that in warm blooded animals, therefore, that milk sugar, after the animal has passed the stage of infancy, might well be eliminated from the diet of those persons in whom there is a familial tendency toward the development of cataract in later life. It has also been found that depancreatized animals have a high incidence of cataract due to the accumulation of blood sugar. There is another belief held by many that excessive accommodation may be one of the causes of cataract due to impaired nutrition of the lens within the eye following its long compression.

[7] Haldane, ''Color Depends on Eye as Well as Wave-Length,'' S.N.L., Nov. 11, 1933.

[8] Erickson, ''Color Blindness,'' S.N.L., 35, 15, 205.

[9] Mitchell, ''Cataract in Rats,'' Jour. Nut., Jan. 1935.

Glaucoma is a disease of the eye which if permitted to continue usually results in blindness which is irreversible by any known therapeutic means. Some authorities hold that glaucoma may be produced by over sympathetic stimulation causing excessive dilation of the pupil and interference with drainage from the anterior chamber of the eye. Others hold that a shallow anterior chamber of the eye may result in the development of glaucoma. *Josephson*[10] holds that a lack of cortin, the secretion of the adrenal cortex, may be the causative factor of this disease and prescribed cortin for the relief or cure of glaucoma. *Josephson* reports a high percentage of cures by following this technique. Since the cortex of the adrenal is parasympathetically controlled and the parasympathetic is somewhat activated by cortin, and that the parasympathetic contracts the pupil, it appears that *Josephson* is on sound physiologic ground in his approach to this very serious eye condition.

The most commonly accepted theory of cause for the phorias apparently does not meet all of the facts set up in its support. It seems probable that nature never makes the two eyes in any one head exactly alike in terms of the exact placement of parts. When it is remembered that any small deviation of the macular area in one eye from the geometrical position of its mate in the other will require that the eye turn in order that the visual axis passing from the point of fixation through the nodal point may coincide with the macula of that eye, it is easy to see that a phoria would result. To the mind of the writer these minute displacements are probably much more important than older theories have held. Certainly, it is true that increasing the act of accommodation or decreasing it will in most instances result in a change in the phoria status of the eyes. But it should be remembered in this connection that to date there are no methods of measuring tendencies toward deviation of the visual axes except by some method which temporarily throws the normal function out-of-gear. Such an ungearing is obviously an abnormal state. Diagnoses of a normal condition based upon measuring an abnormal one, are to the mind of this writer, always open to question. Hence it would seem that all phoria findings should be taken *cum grano salis.*

[10] Josephson, ''Glaucoma and Cortin,'' Chedney Press, 1937.

Another phoria theory held by some is that a phoria is a result of faulty conditioned responses, but since the phoria is not a true deviation of the visual axis such a theory does not seem tenable because in the conditioned responses the act actually takes place, but in the phoria tests, as now used, the event will only take place under the abnormal condition of the test.

Due to the association between the act of accommodation and convergence and the responses set up by the act of accommodation, we do find that in some of the toxic states and particularly those which stimulate the sympathetic, there is an over response of convergence as associated with the act of accommodation. Here the effect found should not be attributed to anything other than its true cause, i.e., a toxic irritation of the nerve cells involved, which is greater than the muscular inhibitory effect produced by the over-active sympathetic. Here are found two variables which make the making of an accurate diagnosis a next to impossible act.

During convalescence from acute illnesses a common finding is a tendency toward exophoria, probably due to partial exhaustion of the adrenals during the acute illness.

Those phorias which are the direct resultant solely of the association of the internal recti with the act of accommodation uniformly respond to irradiation of the eye by some one or more of the deep blue transmitted filters.

Those deviations of the ocular axes from a point of fixation which are objectively determinable, the tropias, were previously accounted for by assuming a paralysis, a paresis or a weakness of the muscle opposite to the direction the eye turns. It is just as tenable to the mind of the writer that the deviation might well have been caused by spastic contraction of the muscles on the side toward which the eye turned. Regardless of these two theories, tropias are known to exist. *Tait* and others hold that the deviation might be the result of a cramp due to wrongly distributed reciprocal tonic innervation and assumes that both the tropias and the phorias have their primary fault in the nervous system. The nervous system in question is the cerebrospinal system. But we also know that over-activity of the sympathetic directly tends to lessen the convergence power. Hence, on theoretical grounds alone there should be a greater tendency to exophoria, exotropia when the sympathetic is over-active. The fact is that we find esotropias in cramp conditions more often than exotropias, so perhaps the reciprocal innervation theory may have to be revised in terms of the autonomic.

The tropia commonly found when the eyes are hyperopic is that of esotropia; when the eyes are myopic the tendency is toward exotropia; and a blind eye almost invariably shows an exotropia. A number of years ago *McCormick* suggested that all esotropias had an accommodative cause and suggested the forced relaxation of the accommodative act for the correction of the tropia. Strangely enough this theory works in a high percentage of cases, regardless of other methods employed. Recently, the amblyopia theory has been advanced as a cause of tropias. This theory is somewhat at variance with the observed fact that a blind eye almost invariably deviates outward, whereas the tropia found in most cases having an amblyopic eye, it being the one that usually deviates, is usually an esotropia. It seems, therefore, that much more study is required in order to arrive at a plausible theory for the causation of tropias, one which will fit all of the facts mentioned above.

Recently there has been advanced the so-called "fatigue theory" for the failure of certain intra- and extra-ocular muscular functions. It seems that the quantitative anemia theory in the brain is much more in conformity with the facts than the fatigue theory. Brain cells and nerve cells are dependent upon a constant and adequate food supply. Failure of such a food supply for a very few seconds results in failure of the nerve cells to function, in fact amounts almost to a paralysis. The amount of food and oxygen supplied to the nerve cells in the brain is determined largely by their vasomotor status, vasoconstriction lessening the food supply, and vasodilation increasing it. But there is another significant physiologic fact which has a bearing here. A quantitatively anemic nerve cell is not only weakened in its power to respond to a stimulus, but it becomes much more *irritable* and tends to over-react to a stimulus. Such over-reaction will exhibit itself in those ocular functions which are controlled by the nerve cells which are for the moment quantitatively anemic. The mechanical and functional status of the upper four thoracic segments of the cord here becomes important because it is these segments which operate through the sympathetic which controls the caliber of the cerebral blood vessels, to say nothing of those in the eyes and the eye muscles themselves. Incidentally, the general systemic status of the sympathetic may also cause quantitative anemia of the brain due to systemic vasoconstriction.

If as the "fatigue" theorists hold, fatigue of a function is the cause of the symptoms exhibited, we should expect to find more symptoms of eye strain and headache of ocular origin in persons whose ocular functions and ability to function are impaired by age. In age fatigue develops at a much more rapid rate than in youth, either due to accumulation of waste products or to a generally low metabolic status. Upon investigation of this problem we find just the reverse to be true. *Fleming*[11] has this to say: "Headache of ocular origin is common in school life and up to middle life, but it is usually a mistake to attribute headache in a person over fifty years of age to eye strain. The headache may occur in younger persons subject to eye strain in the frontal or supra orbital regions, but may also occur at the back of the neck just where it joins the head." Until the fatigue theorists can explain away these facts it seems that the quantitative anemia theory of the author's, voiced above, more nearly fits the clinically observed facts. *Stricker*[12] does find that there is an effect comparable to fatigue which results from continued use of the eyes under degrees of low illumination, but it is not this particular effect which we discussed above.

Recently much work has been done on the subject of night blindness and its relationship to Vitamin A, as found in the pigment layers of the retina. It has been found that a lack of Vitamin A results in very slow dark adaptation or none at all. Many investigators have found that the addition of carotene, the yellow coloring matter which seems to be common to all foods containing Vitamin A, to the diet increases the rate of adaptation of the eye to low degrees of illumination. This effect undoubtedly takes place, but *Rockwell*[13] finds that irradiation of the eyes with a selected band in the red-orange range of the visible spec-

[11] Fleming, "Ocular Symptoms," Clin. Med. Surg., Dec. 31, 1934.
[12] Stricker, "Office Efficiency," Sc. Am., 83.
[13] Rockwell, "Visual Adaptations," Syntonogram, Sept. 1938.

trum will also increase the rate of adaptation to low degrees of illumination.

The writer's theory of one of the causes of amblyopia is that of excessive "synaptic delay." By that the writer means that the accumulated potential difference in the axones to a synapse is not great enough to excite the dendrite to the next cell, and believes that perhaps the theory of *Pasque,* that of condensor action at the synapse, adequately accounts for a large group of cases of amblyopia. Clinically, it has been found that the use of those light frequencies which decrease ionization in the retina are at times effective in so permitting such a build-up of potential difference as to cause an impulse to cross a synapse much in the same manner that an over-charged condensor might rupture its dialectric, thus producing a permanent conduction path. Thus it can be seen that once an impulse jumps a synapse to the dendrite of the next cell a path should become permanently open and would forever after be a conductor of impulses received thereat. Clinically, this is what we find.

There are times in clinical practice when it becomes necessary to increase the amount of accommodation or decrease the amount of accommodation relatively to a given distance and fixation point. This the writer has found to be possible by the utilization of a cross hatched fixation object having clear lines which transmit selected frequencies, separated by areas that are opaque to it. These lines should be about three-tenths of a millimeter wide and the opaque spaces approximately two millimeters square. By such a device interposed between the eye and a series of frequencies such as those transmitted by cobalt glass, there will result an edge diffraction and a separation of light in the eye of the observer into the two extreme frequencies of the visible spectrum. The lines apparently transmitting the high frequency and, due to the diffraction the squares will appear to have transmitted the low frequencies. The chromatic interval between the high frequency and the low frequency transmission of cobalt filters, at about one foot from the nodal point of the eye, is about 1.8 diopters. If the patient fixates such a target at thirteen inches he will require 18 Δ of convergence, for a sixty millimeter interpupillary distance, and if he focuses a neutral color fixation object at that distance he would require 3 D. of accommodation, a ratio of 3 : 1, convergence to accommodation. However, under the conditions set up with the cross hatched fixation object, and the cobalt transmitted frequencies, we now find that with an 18 Δ convergence, a shift of his eyes laterally from a transparent line to an opaque square, a lateral shift of about a millimeter, the accommodation will have to vary through the 1.8 D. if he sees both the line and the apparently red square clearly. By this means it is possible to relax the accommodation to a point behind the convergence point or to force it to a point in front of the convergence point. A series of such treatments usually results in convergence and accommodation being in coincidence in the plane of the fixation object. The effect produced by this treatment is primarily upon the autonomic which controls accommodation, because it is that function which varies under the treatment.

In the foregoing ocular deviations it has been shown that some autonomic anomaly is more or less constantly present, which anomaly may account for the symptoms observed. Simultaneously therewith, we also find certain endocrine malfunctions concomitant with the autonomic deviation. Autonomic deviations may be mass pattern deviations by the sympathetic, or may be more specialized deviations by the parasympathetic. In either event, however, syntony as between the two tends to produce more competent ocular function, both as to the usual functions of seeing, accommodation, convergence, stereopsis, and also in terms of certain diseases of the eye mentioned above.

It follows, therefore, that any attempt to recondition any of the responses of a pair of eyes which does not take into consideration the establishment and maintenance of syntony in the autonomic is foredoomed to *ultimate* failure, regardless of apparent gains at the time. This syntony can be attained by the use of selected light frequencies in the photic range as has been shown heretofore.

CHAPTER XVI

CONCLUSIONS

The following conclusions are the direct outgrowth of the factual and clinical evidence presented herein and represent the author's considered judgment in the evaluation of the data:

1. There exists a closely predictable relationship between light frequency incident into the eyes and their responses.

2. There exists a relationship between light frequency and the rate of growth of cells and tissues, and their rate of cell division.

3. There exists a relationship between the light in the environment and the physical development of the individual.

4. There exists a relationship between light frequency in the eyes and the mass body potentials.

5. There exists a relationship between the light frequency environment and the development of the biotype, modifying the hereditary tendency.

6. There exists relationship between light and light frequency and the action currents leaving the eye toward the brain, these action currents being both quantitatively and qualitatively altered.

7. There exists a relationship between light frequency incident into the eye and the functioning power of the pituitary gland.

8. There exists a relationship between the reproductive cycle and the light frequency environment, probably a quantitative one in respect to the number of individuals of any species.

9. There exists a relationship between the light frequency environment and the dynamic tension present between the two divisions of the autonomic nervous system.

10. There exists a relationship between the light frequency environment and the secretion of hormones by all of the co-acting as well as antagonistic endocrine glands with the pituitary as the "master gland."

11. There exists a relationship which is largely predictable between light frequency environment and the restoration of health following departures from the normal which are still within physiologic limits, particularly those departures which

may be directly influenced by the autonomic or the endocrines toward health.

12. There exists a relationship between light frequency into the eye and the degree of nerve cell irritability thus modifying reflexes.

13. There exists a relationship between light frequency into the eye and bodily health.

14. There exists a relationship between nerve impulses from the eye, due to incident light frequency and the state of tension in the autonomic nervous system.

15. There exists a relationship between light frequency into the eye and either its vitamin A content, or the degree of its adaptation to low degrees of illumination.

16. There exists a relationship between light frequency into the eye and the perception of pain.

17. There exists a relationship between light frequency into the eye and the relative responses of both striped and smooth muscle.

18. Syntony of the autonomic may be produced by light frequency into the eye.

19. The ability to continue to live depends upon syntony of the autonomic in both acute and chronic illnesses, and this attainment of syntony may be aided by light frequency into the eye.

APPENDIX

APPENDIX

Statistical Data

Syntonic Effectivity

A statistical compilation of ocular
anomalies handled by applying the
Syntonic principle

TOTAL NUMBER OF INDIVIDUALS SYNTONIZED 3067
Number of individuals responding to syntonization 2791
Percentage of patients responding to syntonization 90.7

The following 3600 optometric departures from normal were found in the 3076 individuals, some patients having more than one such departure, in addition to those requiring purely wave-optic handling:

	Number	Responded	Effectivity percentage
Phorias, including eso-exo- and hyper-	295	232	78.65
Low blur, break or recovery findings	246	205	83.63
Asthenopia with discomfort	683	634	92.8
Tropias, including eso-exo- and hyper-	103	77	74.85
Amblyopia of undetermined cause	242	185	76.00
Progressive myopia, progress stopped or minus power reduced	68	46	67.65
Headaches of ocular origin	725	629	89.51
Latent hyperopia	64	60	93.74
Color field contractions,—red 47; green 60; blue 65	172	153	88.89
Associated and supportive functions of vision	275	184	66.9
Ocular Reflex or referred pains	144	115	79.79
Opacities, including senile cortical, diabetic, occupations, congenital	425	268	63.59
Optometric departures from normal not classified above	158	112	70.88
TOTAL DEPARTURES FROM NORMAL	3600	2900	80.55

It will be noted from the above that 90.7 percentum of individuals respond to the properly selected Syntonic frequency.

Further it should be noted that 80.55 percentum of optometric departures from normal respond to correctly applied Syntonic frequencies 3076 individuals received 23,993 syntonizations, averaging 7.79 per patient.

211

INDEX